Dortmunder Beiträge zur Entwicklung und Erforschung des Mathematikunterrichts

Band 37

Reihe herausgegeben von
S. Hußmann
M. Nührenbörger
S. Prediger
C. Selter
Dortmund, Deutschland

Eines der zentralen Anliegen der Entwicklung und Erforschung des Mathematikunterrichts stellt die Verbindung von konstruktiven Entwicklungsarbeiten und rekonstruktiven empirischen Analysen der Besonderheiten, Voraussetzungen und Strukturen von Lehr- und Lernprozessen dar. Dieses Wechselspiel findet Ausdruck in der sorgsamen Konzeption von mathematischen Aufgabenformaten und Unterrichtsszenarien und der genauen Analyse dadurch initiierter Lernprozesse.

Die Reihe „Dortmunder Beiträge zur Entwicklung und Erforschung des Mathematikunterrichts" trägt dazu bei, ausgewählte Themen und Charakteristika des Lehrens und Lernens von Mathematik – von der Kita bis zur Hochschule – unter theoretisch vielfältigen Perspektiven besser zu verstehen.

Reihe herausgegeben von
Prof. Dr. Stephan Hußmann
Prof. Dr. Marcus Nührenbörger
Prof. Dr. Susanne Prediger
Prof. Dr. Christoph Selter
Technische Universität Dortmund, Deutschland

Weitere Bände in der Reihe http://www.springer.com/series/12458

Christian Büscher

Mathematical Literacy on Statistical Measures

A Design Research Study

With a foreword by Prof. Dr. Susanne Prediger

 Springer Spektrum

Christian Büscher
Fakultät für Mathematik, IEEM
Technische Universität Dortmund
Dortmund, Germany

Dissertation Technische Universität Dortmund, Fakultät für Mathematik, 2018

Erstgutachterin: Prof. Dr. Susanne Prediger
Zweitgutachter: Prof. Dr. David Pratt
Tag der Disputation: 13.06.2018

Dortmunder Beiträge zur Entwicklung und Erforschung des Mathematikunterrichts
ISBN 978-3-658-23068-5   ISBN 978-3-658-23069-2   (eBook)
https://doi.org/10.1007/978-3-658-23069-2

Library of Congress Control Number: 2018950192

Springer Spektrum

This Springer Spektrum imprint is published by the registered company Springer Fachmedien
Wiesbaden GmbH part of Springer Nature
The registered company address is: Abraham-Lincoln-Str. 46, 65189 Wiesbaden, Germany

Foreword

Statistics education counts as a very important topic for achieving mathematical literacy, as statistics is applied in many everyday and academic situations, inside and outside mathematics classrooms. Although this statement seems obvious, it leaves a lot of open questions. What exactly is the contribution of statistics to mathematical literacy? How can it be seriously promoted, given the fact that in many schools a maximum of two weeks per curriculum year are dedicated to statistics, because arithmetic and algebra take most of the time? How can students' reflections be fostered beyond superficial criticism on manipulations of graphs?

The PhD thesis submitted by Christian Büscher provides theoretically profound and empirically grounded answers to these questions by developing a normative conceptualization of mathematical literacy on statistical measures and by conducting a Design Research study with seventh graders in several iterative design experiment cycles. As a main theoretical contribution, the study develops a normative conceptualization of mathematical literacy with two kinds of goals, *mathematizing goals* (addressing the ability to use mathematical concepts for structuring phenomena in the social and natural world) as well as *reflective goals* (addressing the ability to reflect and evaluate the role mathematics plays in society). But whereas for the mathematizing goals, research in mathematics education has produced a plurality of perspectives and well-elaborated, empirically grounded theories on teaching and learning, the reflective side of mathematical literacy so far remains underdeveloped. This PhD thesis aims to provide a contribution towards closing this *specification gap* by elaborating the notion of *reflective concepts*, which – in analogy to the mathematizing concepts of the mathematizing side of mathematical literacy – need to be developed by learners to be able to critically reflect on mathematics.

His work on the reflective side of mathematical literacy is interwoven with another elaboration, this time on the mathematizing side of mathematical literacy. Statistical measures such as the mean and the median are commonly utilized throughout statistics education research. Christian Büscher substantiates such work by providing a clear conceptualization of the notion of statistical measure. This allows him to describe formal, *general* measures as well as informal, *situative* ones used by learners and to formulate a hypothetical learning trajectory for learners to develop the concept of statistical measure.

One of the main themes running through his work is the trust in learners to develop complex ideas through their own *activities*, if supported by careful design of a teaching-learning arrangement. This is true for the mathematizing side of mathematical literacy as well as for the reflective side. This, however, does not remain a purely theoretical standpoint, but is actualized through his empirical work.

The PhD thesis follows a Design Research program to develop and empirically ground reflective learning opportunities along with the mathematizing learning opportunities which are optimized to let the students develop statistical knowledge. In the empirical core of the thesis, Christian Büscher presents case studies for students' learning pathways towards statistical measures and their mathematizing concepts and the early germs of reflective concepts. These qualitative reconstructions show the complexities of individual processes of guided re-invention and specifically show how mathematizing and reflective activities can and should be intertwined as they stabilize each other in many subtle ways.

Overall, Christian Büscher not only provides a theoretical, normative framework for mathematical literacy and theoretical contributions towards closing the specification gap of reflection. He rather provides empirical proof that learners indeed are able, through their own mathematizing and reflective activities and concepts, to reflect on mathematics while engaging in mathematizing – while providing design principles and concrete design elements that allow educators and researchers alike to initiate such activities.

As a whole, the PhD thesis documents a rich and theoretically deep Design Research project which can substantially contribute to narrowing the gap between the general goals of statistics education and the classroom reality. The thesis accounts for the complexity of reflective statistical learning processes which can be initiated even in a restricted time frame. I hope it will find many readers and future research that builds upon it.

Dortmund, June 2018

Susanne Prediger

Susanne Prediger

Preface

Although in the following pages I argue that meaningful reflection can be carried out while getting work done, there still remains something to be said for taking a step back and looking at a work finished. This is the opportunity to identify all the people that helped along the way.

First, I want to thank Susanne Prediger for her guidance and carefully administered nudges into the right directions, while giving me the room to find my little peculiar niche in mathematics education research. I want to thank Dave Pratt for taking the time for his deep and always constructive commentary on the emergent thesis. Susanne Schnell, who accompanied me on the early steps into statistics education research. The FUNKEN research group, represented through its leadership team of Bernd Ralle, Stephan Hußmann, Jörg Thiele, and Claudia Gärtner for the richness of perspectives and for providing a space for discussing and sharing the woes of Design Research. All members of the working group Prediger/Hußmann and the whole IEEM for providing an invaluable pool of expertise and the willingness to share knowledge; in particular Kristina Penava for being my stochastical companion. The teachers for supporting my search for students, and the students for the willingness to share their thoughts. Martin and Heather Buscher for checking the language of the whole manuscript. And finally, Carina Zindel for all the shared moments of grief and triumph. This wouldn't have been possible without you.

Thank you all.

Table of Contents

List of Figures

List of Tables

1 Introduction

Within the last two decades, statistics education as a field of practice and research has expanded rapidly (Zieffler, Garfield, & Fry, 2018). Increasingly, statistics is recognized as a skill imperative to members of today's society (Engel, 2017). Although implementations vary wildly, statistics has become an integral part of national standards in mathematics education across the world (e.g. for Germany KMK, 2004; for the United States CCSSI, 2018). Statistics is no longer a field of special interest, and effort now is made to include statistics in all levels of education, from primary to tertiary education (Zieffler et al., 2018).

Correspondingly, the body of work in statistics education research has grown enormously (Petocz, Reid, & Gal, 2018; cf. Ben-Zvi & Makar, 2016 and Shaugnessy, 2007). Research has engaged with the field of statistics education from a variety of perspectives, focusing on content (e.g. for authentic statistical practice see Watson, Fitzallen, Fielding-Wells, & Madden, 2018; for uncertainty see Pratt & Kazak, 2018), on specific approaches to teaching (e.g. on the use of computer tools see Biehler, Ben-Zvi, Bakker, & Makar, 2013; on modeling see Lehrer & English, 2018), on level of education (e.g. for teacher education at university level see Frischemeier, 2017), on the use of theory for research (for the specific background theory of inferentialism see Bakker & Derry, 2011), and many more.

The need for research on compact teaching-learning arrangements

Contrary to the increasing knowledge base of statistics education research, statistics education practice seems to not have changed accordingly (Ben-Zvi, Gravemeijer, & Ainley, 2018). In Germany, research indicates that teachers value statistics, yet feel ill-equipped for teaching statistical content in mathematics classrooms (Schumacher, 2017). It seems as if research in statistics education so far has not led to enough results that are easily transformable into classrooms. Recently, this gap has been acknowledged and has led to a call for research to design coherent learning trajectories (Arnold, Confrey, Jones, Lee, & Pfannkuch, 2018) and learning environments useful for educational practice (Ben-Zvi et al., 2018). Thus, statistics education research is in need of more Design Research (Prediger, Gravemeijer, & Confrey, 2015; Prediger & Schnell, 2014).

A key factor to the success of such research is the fit of these designed learning environments to syllabus and classroom reality. In most countries, statistics education is a part of mathematics classrooms (Zieffler et al., 2018). Although some few countries allocate large parts of the mathematics curriculum to statistics (e.g. up to 30 lessons a year in Israel, see Ben-Zvi & Arcavi, 2001), in many

© Springer Fachmedien Wiesbaden GmbH, part of Springer Nature 2018
C. Büscher, *Mathematical Literacy on Statistical Measures*,
Dortmunder Beiträge zur Entwicklung und Erforschung des
Mathematikunterrichts 37, https://doi.org/10.1007/978-3-658-23069-2_1

countries, this leaves very little time for statistics education. For the German context, for example, the syllabus usually plans with about 5-10 lessons per year in middle secondary school. That is why some of the most substantial outcomes of Design Research in statistics education such as the *Nashville Research* (Bakker, 2004) or the *Connections Project* (Aridor & Ben-Zvi, 2017) most undergo a substantial adaptation before fitting into the time allocated in the more restricted time frames of other countries.

This thesis aims to provide a contribution towards closing the gap of applicability. It does so by providing a design of a *compact teaching-learning arrangement* and by researching students' learning processes when interacting with it. The term *teaching-learning arrangement* is used specify a series of problems and accompanying didactic materials given to learners, as well as the role of the teacher and employed methods of teaching. Thus, it presents a part of the "social, cultural, physical, psychological, and pedagogical systems" (Ben-Zvi et al., 2018, p. 476) composing the learning environment. The teaching-learning arrangement needs to be *compact*, as it should achieve the most important goals for statistics education in the little time allocated to it in German classrooms, instead of suggesting large projects with limited hope of implementation. This thesis is situated within the educational context of Germany, but hopefully transferable to other contexts. It aims to provide a theoretically sound and empirically validated teaching-learning arrangement that enables students to step towards becoming informed citizens in a statistics-rich society for any country with little resources for statistics education.

The need for a normative framework for central statistics learning contents

When there is only little time available to learn statistics, one needs to focus on learning contents that are truly important to learn. The current statistics education research landscape is heavily influenced by the construct of Informal Statistical Inference (ISI, Makar & Rubin, 2009). The construct of ISI leans on the practice of professional statisticians, who use probability theory to infer about the behavior of real-world phenomena based on limited observational data. The normative claim surrounding ISI is that inferential statistics is the most important part of statistics, that students should learn statistical inference, and that in school they should engage in statistical inference in an informal way as a preparation for formal statistical inference on college or university level.

Yet some objections can be raised against this approach. From a practical standpoint, the reasoning behind ISI seems to be very complex (cf. Makar, Bakker, & Ben-Zvi, 2011), and thus seems ill-suited for compact teaching-learning arrangements. From a normative standpoint, the complex practice of statistical inference might not constitute an important end-goal of education (Fischer, 2001; to be expanded in Chapter 2). And from a scientific standpoint, a controversial discussion about the actual usefulness of central methods of statistical inference has been going on for almost a century, casting doubts on the appropriateness of an informal approach to inferential statistics (White & Gorard, 2017).

Meanwhile, the practical need for the design of a compact teaching-learning arrangement requires a focus on the centrally important learning contents of statistics. A re-evaluation of the normative foundations of statistics education within the context of German mathematics education is needed to identify relevant statistical concepts and ideas to be developed by learners (to be expanded in Chapters 2 and 3).

The need for an analytical framework for concept development at a micro level

A common demand in mathematics education is to use students' individual conceptions in the development of more formal concepts. Statistics education research has already uncovered some insights into students' individual and informal approaches to data (e.g. Konold et al., 2002; Makar & Confrey, 2003; Makar & Confrey, 2005; Schnell & Büscher, 2015). More insight, however, is needed into possible learning trajectories showing how formal concepts can be developed based on these informal notions (Arnold et al., 2018).

Describing such learning trajectories requires a framework that enables both, specifying the learning content from a subject-matter perspective, as well as describing students' individual conceptions and sense-making from the learners' perspective. To specify the connections between learners' conceptions and statistical concepts, such a framework also needs to be usable as a language of description. This thesis constructs a framework that enables the construction of a hypothetical learning trajectory spanning from learners' individual conceptions towards the highest goals of statistics education (in Chapter 4). It then elaborates the same framework into an analytical framework to provide empirical insights into students' concept development at a micro level (in Chapters 7 and 8).

The need for a descriptive framework for students' reflections in learning processes

One central part of mathematical literacy is the ability to reflect on the role of mathematics in the world. The need for reflection already has a well-established tradition in mathematics education research (e.g. Skovsmose, 1994; Fischer, 2001; Wille, 1995; Lengnink & Peschek, 2001). Such demands for reflection in mathematics education, however, often stay general in nature, leaving only little hints how to implement reflection in the classroom. Although some more concrete proposals do exist (e.g. Skovsmose, 1998; Prediger, 2005b), detailed empirical accounts of reflection are hard to find.

One possible reason might be a missing framework allowing to describe students' informal reflections situated in learning processes. Reflection is mostly specified on a normative level as a general learning aim of mathematics education (as described thoroughly in Chapter 2). Beyond this, a descriptive framework is required that allows to identify students' reflections in learning processes so that a detailed account of students' reflections can be given. This is a necessary step that could then allow to formulate learning trajectories from students' reflections towards the general aims concerning reflection in mathematics education. This thesis provides such a conceptualization of learners' situated reflections (in Chapter 4), which allows an empirical reconstruction of learners' reflections (in Chapter 8).

Structuring this thesis

This thesis pursues the design of a compact teaching-learning arrangement for statistics in two parts: the theoretical part (Chapters 1 to 5) provides the normative and prescriptive specification of the learning content of statistical measures; the empirical part (Chapters 6 to 9) introduces the compact design of a teaching-learning arrangement concerning statistical measures and describes the empirical thesis evaluating its design principles and elements, as well as students' learning processes.

Chapter 2 introduces the normative framework of *mathematical literacy in statistics*, which gets split into the two sides of the *mathematizing side* and the *reflective side* of mathematical literacy in statistics. This framework then is used to identify general learning *aims* as well as specific learning *goals*. Whereas for the mathematizing side the aims and goals can be easily specified, an existing specification *gap* hinders the specification of the goals of the reflective side of mathematical literacy in statistics. Within this theoretical chapter, the concept of measure is identified as a central learning content.

Chapter 3, however, shows that the concept of measure is rarely explicitly focused by statistics education research. The chapter provides a conceptualization of measure through the constructs of *formal characteristics, features of the data*, and *aspects of phenomena* as well as through the distinction between *situative* and *general measures*. The concept of measure then is linked to the construct of Informal Statistical Inference.

Chapter 4 provides the learning-theoretical background of this thesis. This background is used to link the two types of learning goals, mastering activities and developing concepts. It also provides a theoretical foundation for students' informal and intuitive reflections through the construct of *situative reflective concepts*. Regarding the mathematizing side of mathematical literacy in statistics, a hypothetical learning trajectory shows the connection from learners' situative reasoning towards the learning goals. Regarding the reflective side however, this thesis identifies an existing *realization gap* that hinders the a priori formulation of such a hypothetical learning trajectory for reflective concepts.

Chapter 5 summarizes the theoretical work and introduces the empirical research questions for the empirical part of this thesis.

Chapter 6 introduces the methodological frameworks of Topic-Specific Didactical Design Research Reconstruction that support the research of this thesis. It also describes the methods of data collection and analysis.

Chapter 7 identifies several design principles that can support the development of the learning goals of the mathematizing side and introduces the design of a compact teaching-learning arrangement for the concept of measure. The learning processes of two pairs of students are analyzed on a micro-level revealing their rich conceptual networks. This leads this thesis to insights about the role of specific design principles and design elements of the teaching-learning arrangement.

Chapter 8 revises the design principles based on the results of Chapter 7 to increase the focus on the reflective side of mathematical literacy in statistics. By reconstructing students' conceptual networks for two additional pairs of students, this chapter identifies several reflective concepts in a contribution to close the specification gap identified earlier. It also provides insights into students' situative reflective concepts and the role of element of the teaching-learning arrangement for students' reflections.

Chapter 9 summarizes this thesis by revisiting the central theoretical as well as empirical results. A discussion of the studies' limitations and an outlook on further necessary research conclude this thesis.

As point of departure for the development of a compact teaching-learning arrangement, this PhD-Thesis begins by investigating the construct of mathematical literacy as a normative framework for specifying central statistics learning contents

2 Mathematical literacy in statistics

The previous chapter illustrated the need for careful consideration of learning content for a compact teaching-learning arrangement in statistics. A re-evaluation of the general aims (*why* to learn) as well as more specific goals (*what* to learn) of a statistical education is in order. Other approaches to a normative framework for statistics education emphasize the differences between statistics and mathematics (Cobb & Moore, 1997; delMas, 2004) and draw on the practice of statisticians for providing aims and goals of statistics education (e.g. Pfannkuch & Wild, 1999; Makar & Rubin, 2009). This study, however, emphasizes the similarities between statistics and mathematics. This allows this study to draw on a wide range of normative frameworks for education already established by researchers, philosophers, and didacticians of mathematics.

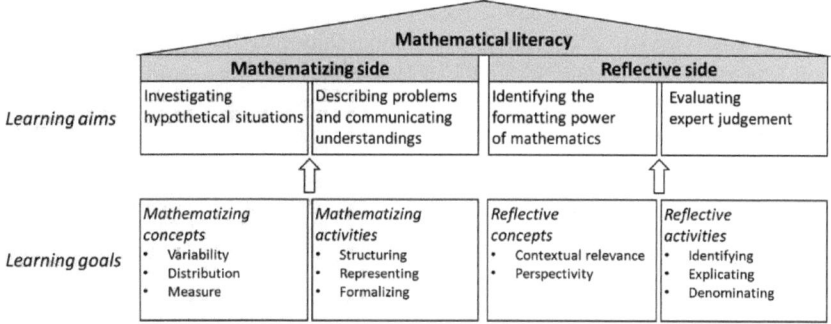

Fig. 2.1: Advance organizer for the contents of this chapter

Figure 2.1 provides an advance organizer for the contents of this chapter. The point of departure of this study is the construct of mathematical literacy, which consists of the two sides of mathematizing and reflecting (Section 2.1). After explicating the connections between mathematical literacy and statistics (Section 2.2), the sides of mathematical literacy are investigated separately. Each side of mathematical literacy contains its own general aims (Sections 2.3.1 and 2.4.1), answering the question *why* to attain mathematical literacy. The question *what* to learn for each side of mathematical literacy is answered by providing two different kinds of goals: the mastering of specific *activities* (Sections 2.3.2 and 2.4.2) and the development of specific *concepts* (Sections 2.3.3 and 2.4.3). Reconnecting to the task of designing a teaching-learning arrangement for statistics, the concept of measure is introduced as a learning content that can serve as a focus for both sides of mathematical literacy (Section 2.5).

© Springer Fachmedien Wiesbaden GmbH, part of Springer Nature 2018
C. Büscher, *Mathematical Literacy on Statistical Measures*,
Dortmunder Beiträge zur Entwicklung und Erforschung des
Mathematikunterrichts 37, https://doi.org/10.1007/978-3-658-23069-2_2

2.1 The mathematizing and reflective sides of mathematical literacy

The question of the general aims and purposes of education in mathematics has a long-standing tradition in mathematics education research, which is connected to the German constructs of *Bildung* and *Allgemeinbildung* (cf. Heymann, 2010 for an English overview on the discourse in Germany). Following the line established by Winter's (1996) 'basic experiences', answers often regard the importance of the instrumental use of mathematics in real-world application (e.g. de Lange, 1997), the significance of an enormous deductive theory as a human cultural achievement (e.g. Hardy 1940/2005), or its role as expression of a unique mode of human thought (e.g. Pólya, 1945). A construct that has become influential internationally is the construct of *mathematical literacy* (Niss & Jablonka, 2014). Many different meanings are associated with this term (Jablonka, 2003) which is consequently lacking consensus in mathematics education research (Niss & Jablonka, 2014). However, due to its strong influence on national curricula, the definition given in the PISA 2015 Mathematics Framework (OECD, 2017) provides a starting point for further investigation:

"Mathematical literacy is an individual's capacity to formulate, employ and interpret mathematics in a variety of contexts. It includes reasoning mathematically and using mathematical concepts, procedures, facts and tools to describe, explain and predict phenomena. It assists individuals to recognize the role that mathematics plays in the world and to make well-founded judgements and decisions needed by constructive, engaged and reflective citizens" (OECD, 2017, p. 67).

Two different sides of mathematical literacy can be identified based on this definition: (a) an individuals' capacity to *use* mathematics for understanding phenomena of the world, and (b) an individuals' capacity to *evaluate* the role of mathematics in society (see also Büscher & Prediger, submitted).

The first side of mathematical literacy has its roots in the works of Freudenthal (1973; 1983a; 1991). To Freudenthal, the formal mathematical system of definitions, theorems, and proofs does not constitute an aim of mathematics education in itself. Rather, he regards formal mathematics as only the end product of extensive activity of *mathematizing*, and putting the product of formal mathematics before the activity of mathematizing constitutes an 'anti-didactical inversion': "Mathematics has arisen and arises through mathematizing" (Freudenthal, 1991, p. 66). Therefore, instead of the end product of mathematics, the activity of mathematizing has to be the focus of mathematics education.

Central to his work is the idea of a *phenomenology* of mathematical structures (Freudenthal, 1983a). Mathematics is not seen as an abstract body of knowledge disconnected from the world, but rather as having its roots in everyday phenomena:

"Our mathematical concepts, structures, ideas have been invented as tools to organize the phenomena of the physical, social and mental world. *Phenomenology* of a mathematical concept, structure, or idea means describing it in its relation to the phenomena for which it was created, and to which it has been extended in the learning process of mankind [...]" (Freudenthal, 1983a, p. ix).

One conclusion is that mathematics education has to focus on phenomena, allowing learners to develop mathematics through mathematizing and to see the value of mathematics in structuring phenomena of the world. This focus on phenomena has since been adopted and thoroughly expanded on in the reality principle of the research program of RME Theory (van den Heuvel-Panhuizen, 2001). Apart from RME Theory, the general principle of focusing on the investigation of phenomena to create understanding also is a common theme in mathematics education research, with examples such as the exemplary and genetic principles of Wagenschein (1968/2010) or the focus on using statistics to explore phenomena rather than confirm pre-formed hypotheses by Tukey (1977). To emphasize these conceptual roots, the side of mathematical literacy that concerns the use of mathematics to investigate phenomena here is conceptualized as the *mathematizing side* of mathematical literacy.

The second side of mathematical literacy concerns the evaluation of the role mathematics plays in the world. As Jablonka (2003) points out, this relationship can also be conceptualized in many different ways. Mathematics can be conceptualized as an expression of culture, and learners should learn to find the mathematics in the artifacts of local culture (Bishop, 1988). It can be conceptualized as a certain power acting on society, which learners should learn to identify in order to initiate social change (Skovsmose, 2005). It can also be conceptualized as a mediator between individual and society, and learners should learn how mathematics regulates this relationship (Fischer, 1988; Fischer, 1993a; Lengnink & Peschek, 2001; Wille, 2008).

A common theme for this side is the demand to not only use mathematics, but to *reflect* on the use of mathematics (see Skovsmose, 1994; Fischer, 2001; Prediger, 2005b). Again, different conceptualizations of the notion of reflection exist in mathematics education research, calling for further clarification (see below). Nevertheless, the side of mathematical literacy that concerns the evaluation of the role of mathematics in the world here is conceptualized as the *reflective side* of mathematical literacy.

The reflective side of mathematical literacy brings a critical element to the conception of mathematical literacy. Jablonka (2003) emphasizes that the aim of critical evaluation forms its own distinct knowledge base:

"The ability to evaluate critically can neither be considered as mathematical, nor automatically follows from a high level of mathematical knowledge" (Jablonka, 2003, p. 98).

Thus, the mathematizing and reflective sides of mathematical literacy each need to be addressed specifically, and their knowledge bases need to be identified separately.

As sketched earlier however, there exist certain views of statistics and mathematics that emphasize the differences between these two domains. As such, the use of the construct of mathematical literacy for this research in statistics education requires some additional explanation. This is presented in the next section.

2.2 Mathematical literacy in statistics

In Germany, statistics education is commonly considered a part of mathematics education, and thus belongs to the domain of mathematical literacy. Within statistics education, however, there exists a common distinction between mathematics and statistics. This distinction, however, should not prohibit the use of the rich body of knowledge surrounding the construct of mathematical literacy for statistics education research. Instead, this section argues that authors emphasizing differences between statistics and mathematics tend to hold a certain view on mathematics that does not conform to the idea of mathematizing outlined earlier. Section 2.2.1 explicates these views and shows how the mathematizing can indeed be considered a central part of statistics. Section 2.2.2 then proceeds to explicate connections between the constructs of statistical literacy and mathematical literacy.

2.2.1 Mathematizing as a statistical activity

Statistics education and statistics education research are relatively new disciplines, emerging only in the 1980s as distinct fields (Zieffler et al., 2018; Petocz et al., 2018). In order to emancipate the emerging fields from mathematics education and mathematics education research, the differences between statistics and mathematics were often emphasized. As an example, one point commonly made refers to a different role of context for statistics and mathematics, summed up by Cobb and Moore (1997): "In mathematics, context obscures structure" (ibid., p. 803), whereas "in data analysis, context provides meaning" (ibid., p. 803). Cobb and Moore see mathematics as being concerned only with numbers and structure, not with context and content. The actual *use* of mathematics only concerns applied disciplines such as statistics and physics:

"[Mathematical understanding] is not even the most helpful kind [of understanding] in most disciplines that employ mathematics, where understanding of the target phenomena and core concepts of discipline take precedence" (Cobb & Moore, 1997, p. 815).

Thus, understanding phenomena is not a part of mathematics itself. This point is also made by delMas (2004), who argues that "both language and mathematics can be considered abstract artifacts of human intellect and culture" (delMas, 2004, p. 83). DelMas points out that mathematical objects do not gain meaning from context, but from their place in a deductive system of formal mathematics. Thus, he emphasizes the abstract nature of mathematical entities over the practical real-world application of statistics. Recently, Wild, Utts, and Horton (2018) again stressed this point:

"And whereas in mathematics, mathematical structures can exist and be of enormous interest for their own sake, in statistics, mathematical structures are merely a means to an end" (Wild et al., 2018).

Thus, Wild et al. conceive of mathematical structures as abstract objects independent from the real world. Mathematics then is the science of these abstract objects, and is interested only in the relationships between these theoretical entities.

In the philosophy of mathematics, such views on mathematics that emphasize the disconnectedness of abstract mathematical objects from the real world and their place in the axiomatic system of mathematics are counted to the *formalist* views on mathematics (cf. Lakatos, 1976). As outlined in Section 2.1 however, there are also other distinct views on mathematics which are much more influential for modern mathematics education and mathematics education research. In particular, the view of mathematics as an activity worked out by Freudenthal (1991) refutes large parts of the above views on mathematics. According to Freudenthal, the abstract axiomatic body of mathematics is merely the end product of the activity of mathematizing, which should be held as the most important part of mathematics for education. The activity of mathematizing again consists of investigating and structuring phenomena of the real world. From a perspective of mathematics as mathematizing, the split between 'abstract' mathematics and 'practical' statistics cannot be sustained.

However, a perspective of mathematics as mathematizing can help to find similarities and connections between mathematics and statistics. While characterizing statistics, Wild et al. (2018) also identify aspects of statistics that fit to the activity of mathematizing:

"The mission of statistics education is to provide conceptual frameworks (structured ways of thinking) and practical skills to better equip our students for their future lives in a fast-changing world" (ibid., p. 6).

"Statistics is a meta-discipline in that it *thinks about how to think* about turning data into real-world insights. Statistics as a meta-discipline advances when the methodological lessons and principles from a particular piece of work are abstracted and incorporated into a theoretical scaffold that enables them to be used on many other problems in many other places" (ibid., p. 7).

The aim of statistics is to provide conceptual frameworks for living in a fast-changing world. Mathematizing is an activity of structuring phenomena through mathematical concepts in order to understand them. Statistics is a meta-discipline that advances by reflecting on and abstracting from methodological lessons. Mathematizing is an activity of ordering and reflecting on phenomena, resulting in the higher-order mathematical concepts of the meta-discipline of mathematics. Statistics uses theoretical scaffolds to investigate new problems in new places. Mathematizing is an activity that can draw on a canon of general mathematical concepts for understanding new phenomena.

Thus, under a perspective of mathematics as an activity, the similarities between mathematics and statistics begin to reappear. Both disciplines aim at understanding phenomena, and both disciplines advance by reflecting on processes of understanding phenomena. This is captured by the activity of mathematizing. Thus, mathematizing also is a statistical activity. Statistics education research should correspondingly draw on the vast body of knowledge concerning mathematizing that was produced by mathematics education research.

2.2.2 Mathematical literacy as an integrating framework for statistical literacy and reasoning

Another area of similarity concerns the constructs of mathematical literacy and statistical literacy. Both constructs conceptualize an ability needed by citizens in modern society. However, whereas the former also includes the ability to *use* mathematics for understanding phenomena, the latter mostly refers to *interpreting* given statistics. Gal (2002) defines statistical literacy by two interrelated components:

"(a)   people's ability to interpret and critically evaluate statistical information, data-related arguments, or stochastic phenomena, which they may encounter in diverse contexts, and when relevant

(b)   their ability to discuss or communicate their reactions to such statistical information, such as their understanding of the meaning of the information, their opinions about the implications of this information, or their concerns regarding the acceptability of given conclusions." (Gal, 2002, p. 2/3)

Thus, the development of statistical literacy is conceptually separated from the development of statistical reasoning – broadly speaking, the ability to use statistics to understand phenomena. In practice and many school curricula however, both constructs appear to be intricately linked. It appears that statistical literacy could be developed while developing statistical reasoning, leading Watson, Fitzallen, Fielding-Wells, and Madden (2018) to question the need of separate developments of statistical literacy and reasoning.

Mathematical literacy, however, concerns both, using and reflecting on mathematics. Thus, the construct of mathematical literacy could serve as an integrating framework for these two facets, illustrating how statistical literacy and statistical reasoning belong to the same general competence. Additionally, the reflective side of mathematical literacy has already been elaborated through different philosophical perspectives, resulting in a pronounced view on the role of mathematics in the world (see Section 2.4). Statistics education research should capitalize on this work to clarify what it means to be able to "critically evaluate statistical information" (Gal, 2002, p.2).

Thus, this study uses the construct of mathematical literacy as an orienting framework, because it provides a wider and at the same time more elaborated framework than the constructs of statistical literacy and statistical reasoning would do without this framework. To indicate the concern with literacy regarding statistics, but acknowledging the conceptual differences between mathematical literacy and statistical literacy, this study from here on refers to *mathematical literacy in statistics*.

2.3 Aims and goals of the mathematizing side of mathematical literacy

As established in Section 2.1, the mathematizing side of mathematical literacy concerns the individual's ability to mathematize phenomena of the world. However, mathematics in general, and statistics in particular, is a vast field of study, and such a broad definition alone cannot serve as a foundation for justifying the choice of specific learning content. As Arnold et al. (2018) point out, teaching-learning arrangements have to be embedded in learning trajectories towards specific learning *goals*. These goals can comprise the development of a specific concept or procedure, or the mastering of a specific activity. Reaching these specific goals, however, does not provide the ultimate rationale for learning. There also exist goals of a higher order, such as enculturating students into the practice of professional statisticians (e.g. Ben-Zvi, 2004). Such higher-order goals cannot be attained directly, but instead serve as orienting normative learning *aims* that provide a justification for specific learning *goals* such as developing a specific concept. Learning aims describe *why* one needs to learn, whereas learning goals describe *what* one needs to learn.

To provide such a framework for justifying learning contents, the study first identifies two learning aims of the mathematizing side of mathematical literacy, which here are called *mathematizing* aims (2.3.1). These mathematizing aims then serve as a foundation to identify specific *mathematizing goals* that should be reached, namely a list of specific activities to be mastered (2.3.2) and specific concepts to be developed (2.3.3).

2.3.1 Two mathematizing aims

To identify the general mathematizing aims, this study draws on approaches from the philosophy of mathematics. First, the utility of mathematics as means to investigate hypothetical situations is emphasized. Second, the general nature of mathematics as a means and system of communication is illustrated.

Investigating hypothetical situations

Although the instrumental usefulness of mathematics for real-life situations is often emphasized (e.g. in the PISA definition of mathematical literacy), Danish researcher in mathematics education Ole Skovsmose (2005) notes that mathematics is not confined to describing only real situations:

"By means of mathematics, it is possible to establish a space of hypothetical situations in the form of (technological) alternatives to a present situation. [...] By means of mathematics we seem able to investigate particular details of a not-yet-realized design. Thus, mathematics constitutes an important instrument for carrying out detailed thought experiments" (Skovsmose, 2005, p. 87).

The importance of mathematics thus lies in the possibility of investigating hypothetical situations. One kind of hypothetical situation could be created by using mathematics to predict the hypothetical behavior of actual phenomena – a competence already part of the PISA definition of mathematical literacy. The use of mathematics to investigate hypothetical situations, however, transcends the pure prediction of *actual* phenomena. Mathematics can also be used to describe, explain, and predict *hypothetical* phenomena.

One example could be models of the economy commonly used in political decision-making. Such models simulate the complex phenomenon of the whole national economy through various parameters concerning allocation of workforce, behavior of consumers, and effects of taxation. The model, however, is not bound to a concrete reality. Instead, the model can then be investigated, parameters can be systematically varied, and possible consequences can be explored. The predicted effects of such a model could then be used to justify decisions by policy-makers whether to try to change the current real situation. This is only made possible because mathematical models do not need to refer to 'real' facts, but allow insights into hypothetical situations.

Such a purely instrumental use of mathematical modeling, however, is not the main point made by Skovsmose (1994; 2005). In his critical theory, the investigation of hypothetical situations can provide mathematics education with an element of empowerment. Rather than focusing on society's status quo, one can imagine, via mathematics, a different world. And students should learn to do just that: using mathematics, they should explore the dependencies and power relations within real-world situations, and by investigating hypothetical situations, they should become aware of the possibility for change.

Describing problems and communicating understandings

A possible mathematizing aim commonly proclaimed is the ability to use mathematics to solve real-world problems. For this aim, mathematics education tries to teach students the problem-solving skills employed by experts, and development of expert reasoning becomes the main goal of education (e.g. for statistics Pfannkuch & Wild, 1998; Makar & Rubin, 2009).

Austrian philosopher and researcher in mathematics education Roland Fischer vehemently disagrees:

"Regarding today's focus on 'problem-solving ability' as the aim of (school) education, one has to say: in post-school praxis, nobody solves problems with (school-) subject-specific means" (Fischer, 2013, p. 335 own translation).

On the one hand, mathematical procedures (beyond elementary mathematics) are no part of problem-solving processes of daily life. On the other hand, Fischer (2001; 2013) also questions the need for education to produce citizens who are able to perform tasks that actually have their own dedicated field of experts. Members of society do not themselves need to be able to build houses or to carry out statistical investigations. Rather, citizens are surrounded by experts who can be drawn upon in decision-making. Instead of personally carrying out the task, one just needs to be able to provide experts with the necessary information. The experts can then use their expertise and provide an answer.

This results in a shift of the mathematizing aims from problem *solving* to problem *description* (Fischer, 1993). This does not lessen the importance of learning mathematics, but only emphasizes an important aspect differing from the 'tool-box' view of mathematics:

"Mathematics regulates the social (corporate) life of people in various respects. It is simultaneously a means and a system of communication. This establishes a connection between the individual and society" (Fischer, 1988, p. 28).

As a *system* of communication, mathematics provides a well-defined vocabulary and a set of binding rules for its use. As a *means* of communication, it allows individuals to utilize this shared system of communication to make their understanding of problems relatable to others. Mathematics enables interpersonal communication by providing an intersubjective system of communication.

This role of mathematics in facilitating communication, however, is not restricted to an individual's ability to personally communicate with others only. Mathematics also fulfills an important role in allowing societies to grow: as they become larger, the organizational functions of society are less well suited to personal communication – societies become too large for everyone to know everyone (Fischer, 1986). Communication can no longer rely on shared experiences between individuals who know each other. Here, mathematics can provide a formal system of communication to exchange information, allowing the organization of large societies.

This role of facilitating communication shows to be especially important to the specific subject of statistics. In the early history of statistics, Quetelet (1842/1994) worked on just such a theory to enable communication about otherwise individual and intuitive judgements:

"When a physician is called to examine the body of a lifeless infant, and when, in a legal inquiry, he, *from simple inspection*, establishes the presumed age of the child, it is evident that he cannot but impose his judgement on those who read the inquiry,, however, erroneous it may otherwise be, *since there are no elements existing for the verification of it*" (Quetelet, 1842/1994, p. 72, emphasis added).

Even if the physician's judgement is sound, without statistical theory he is entirely unable to explain his reasoning, because it is formed through experience and practice. This also makes it impossible to criticize his judgement. By introducing the conception of the 'average man', Quetelet (1842/1994) constructs a theory that allows to communicate such common-sense reasoning that deals with the height people tend to have for their certain age.

By analyzing the historical impact of the development of statistical theory on scientific practice and systems of commerce throughout history, Porter (1995) shows how advances in statistical theory allowed society to grow. Quantification and measurement allowed personal scientific knowledge gained through individual scientific practice to be transferred between laboratories and thus assisted in the creation of a community of science. And, the standardization of measurements enabled the creation of a shared system of commerce and thus the groundwork for large economic systems.

The relationship between statistics and communication is also emphasized by Abelson (1995), who takes account of the role of statistics in scientific argumentation:

"[…] I have arrived at the theme, that the purpose of statistics is to organize a useful argument from quantitative evidence, using a form of principled rhetoric" (Abelson, 1995, p. xiii).

Statistics needs to communicate its findings to an audience. This is not only a simple additional task, but rather an integral part of all its activities:

"Data analysis should not be pointlessly formal. It should make an interesting claim; it should tell a story that an informed audience will care about, and it should do so by intelligent interpretation of appropriate evidence from empirical measurements or observations" (Abelson, 1995, p. 2).

This shows how statistics creates both, system and means of argumentation. It is a mathematizing aim of mathematical literacy for students to engage in this system by describing problems and communicating their understandings.

2.3.2 The mathematizing goal of mastering mathematizing activities

Just like the mathematizing aims have different formulations based on the different roles of mathematics, so do the mathematizing goals based on different characterizations of mathematics. Two types of characterizations can broadly be identified: the views of mathematics as a set of specific *activities* and as a set of specific *concepts*. They differ mostly in their emphasis on what is to be learned. One example for the former view is the goal of achieving competence in a subject outlined by Klieme et al. (2007), who define competence as "a disposition that enables individuals to solve specific types of problems, that is to overcome concrete challenging situations" (Klieme et al., 2007, p. 72 own translation). Overcoming such situations requires individuals to *act*, and proponents of such a view for mathematics education emphasize the importance of certain 'mathe-matical' activities. Proponents of the latter view emphasize the importance of mathematical 'core concepts' or 'fundamental ideas' (e.g. Burrill & Biehler, 2011). Both views are interconnected and belong to the mathematizing side of mathematical literacy, as mathematizing consists of the *activity* of organizing phenomena through mathematical *concepts*. This duality of views manifests in theory by drawing connections between conceptual and procedural knowledge, and in educational practice by curricula distinguishing between 'general' and 'content-related' competencies (e.g. KMK, 2004). Yet for analytical purposes, the two types of goals are treated separately here, starting with the goal of mas-tering mathematizing activities.

A view on mathematics that characterizes mathematics by identifying basic mathematical activities sees mathematics as a "concrete formation of general thought-acts" (Prediger, 2004, p. 62, own translation). Different authors sub-scribing to such a view, however, often produce different lists of activities to be considered the 'basic' mathematical activities. Freudenthal (1991), identifying the smaller activities making up the meta-activity of mathematizing, lists activi-ties such as generalizing, organizing, and structuring. The mathematician Mac Lane (1986) identifies the roots of mathematical concepts in 'human cultural activities' such as counting, comparing, and arguing. From a perspective of ethnomathematics, Bishop (1988) identifies a small set of mathematical activi-ties he holds to be invariant across cultures such as counting, measuring, and playing. From a didactical perspective, Fischer and Malle (1985/2004) produce an extensive list of general 'activities' such as generating examples, describing formally, and generalizing. Wille (2000) lists 'general thought-acts' such as abstracting, formalizing, and systematizing.

These lists of mathematical activities show similarities as well as differ-ences. 'Formalizing' is an activity included by both Freudenthal and Wille, 'counting' by Mac Lane and Bishop; the activities listed by Mac Lane and Bish-op are more focused on specific mathematical concepts (counting, measuring, estimating), whereas the activities listed by Malle and Fischer focus more on content-independent general heurisms (generating examples, systematizing). Because this study is not concerned with the comparison of conceptions of mathematics, but rather with the identification of suitable goals for a teaching-learning arrangement in statistics, no further comparison between the lists of mathematical activities is made.

In light of this variety of possible activities as mathematizing goals, a selec-tion needs to be made. For the purpose of this study focusing on statistics in Grade 7, activities such as counting, locating, and axiomatizing do not seem very relevant. And instead of exhaustive and detailed enumerations of activities, a compact teaching-learning arrangement also needs a compact list of activities. As such, this study chooses three *mathematizing activities* to be mastered: structuring, representing, and formalizing. To limit the discourse about the nature of mathematics, these mathematizing activities are illustrated by examples rather than closely defined (Tab. 2.1).

Tab. 2.1: The three mathematizing activities

Mathematizing activity	Examples
Structuring	Identifying patterns
	Identifying relevant parts of phenomena
	Explaining differences between phenomena
Representing	Creating symbols
	Drawing sketches
	Taking procedures as objects
Formalizing	Finding procedures for calculation
	Designating properties

These three mathematizing activities are not intended to provide a complete list of mathematizing goals, but constitute a practical selection of activities to focus on. However, these three mathematizing activities can already serve as mathematizing goals relevant to the mathematizing aims. An investigation into a hypothetical situation is carried out by structuring highly formalized phenomena represented through diagrams or models, and describing and communicating in the system of mathematics requires the formalization and representation of the structure of phenomena.

As already shown earlier, the identification of suitable mathematizing activities is only one possible answer to the question of what constitutes the mathematizing goals. Another answer focuses on relevant concepts to be developed by the learner.

2.3.3 The mathematizing goal of developing mathematizing concepts

Another approach to the identification of relevant mathematizing goals is to provide a list of concepts to be learned. By focusing on *statistical* concepts, this also provides an opportunity to already provide statistic-specific goals to the general mathematizing aims. Within statistics education research, several of such lists exists. Following the footsteps of Freudenthal, Bakker (2004) engages in a historical phenomenology to identify average, sampling, median, distribution, and graphs as statistical concepts relevant especially to the early stages of statistics education. Makar, Bakker, and Ben-Zvi (2011) identify variability, distribution, inference, and sampling as relevant statistical concepts to be used in greater statistical activity. Burrill and Biehler (2011) identify 'fundamental statistical ideas', listing data, variation, distribution, representation, association, probability, and sampling and inference.

This demonstrates a loose consensus on the importance of some statistical concepts (e.g. distribution and variability), yet not a definite list of statistical concepts to be developed. Additionally, although some clarifications of specific concepts in expert practice do exist (e.g. Gould, 2004; Wild, 2006), work remains in further expanding on these concepts and their links to each other. Relevant statistical concepts for the mathematizing aims are identified as follows. This list, however, should not be taken as a complete and sufficient list of statistical concepts, but rather as a list of potential mathematizing goals that can inform the design of a teaching-learning arrangement.

Variability

Moore (1990) sees uncertainty as the central idea of statistics, caused by an 'omnipresence of variability' (Cobb & Moore, 1997). The fact that things are never exactly the same is a fundamental phenomenon of everyday life: people vary in height, temperatures vary from day to day, and coin tosses produce different outcomes. Yet some kind of pattern can often be observed: men generally are taller than women, warm weather is typical for summer, and heads and tails will balance out in the long run. Statistics is the science of understanding this phenomenon of variation, of classifying different kinds of variation, and of finding ways to approximate certainty in an uncertain world.

The *phenomenon* of variation observed in data can be abstracted as the characteristic of data to have a 'propensity for change' (Reading & Shaughnessy, 2004, p. 202), an important part of the *concept* of variability. Statisticians use their knowledge of variability to drive their statistical investigations (Gould, 2004). An (overly simplified) aim of these investigations is to find features in the variation of the data that hint at some stable aspect of the data-generating process or underlying population, and to distinguish these features from variation introduced by irrelevant factors such as artifacts produced through the data gathering process or simple random chance. This distinction is often framed in terms of distinguishing between the 'noise around a signal' (Konold & Pollatsek, 2002), between 'pattern and variation' (Schnell, 2013), or between 'explained and unexplained variation' (Wild, 2006). As Gould (2004) points out, however, both sides of this distinction are important, and an isolated focus on just the signal or pattern is an expression of a 'naïve' conception of variability.

Distribution

Closely connected to the concept of variability is the concept of distribution: it is the "lens through which we [statisticians] view variation in data" (Wild, 2006, p. 13). To characterize the type of variation observed in data, statisticians match the empirical, observed distributions to theoretical, abstract distributions. This provides them with standardized models that allow the use of statistical procedures to generate possible new insights into the phenomenon under investigation – theoretical insights which still need to be compared to the actual phenomenon at hand (Wild, 2006). Looking at data with the concept of distribution requires statistical investigators to hold an 'aggregate view' on data (Konold, Higgins, Russell, & Khalil, 2015). In such a view, data should not be seen as a simple collection of individual values, but rather as an entity on its own with emergent characteristics. The aggregate shows characteristics that cannot be attributed to any single datum it is made of, such as the overall center or spread of the data. Characteristics such as center, spread, density, and skewness are not only properties of empirical data, but also of theoretical distributions (Bakker & Gravemeijer, 2004). As such, holding an aggregate view on data thus is an important part of the concept of distribution.

Measure

A statistical concept often referred to only implicitly is the concept of *measure*. Although some studies have focused on specific measures, mostly mean (e.g. Konold & Pollatsek, 2002) and median (e.g. Bakker & Gravemeijer, 2006), only little attention is given to the general concept of measure. This is surprising, seeing the ubiquitous nature of measures: large swaths of empirical research use mean, standard deviation, p-value, et al. to formulate results; political arguments are fought out via use of statistical measures; and measures are defining characteristics of theoretical distributions. Chapter 4 shows that the concept of measure can be a central concept for the development of mathematical literacy. First, however, the concept of measure is in need of a clear conceptualization and of a clarification of the relationship to the concepts of distribution and variability. This is the focus of Chapter 3.

Development of Mathematizing concepts as a mathematizing goal

The concepts of variability, distribution, and measure are necessary for reaching the mathematizing aims. The investigation of hypothetical situations through statistics is done by describing, predicting, and explaining the variability of hypothetical phenomena; it is also the subject of communication pursued through statistics. Distributions and measures provide the means of simulating situations and the means of communication. Thus, the development of these concepts constitutes a mathematizing goal for reaching the mathematizing aims; the concepts of variability, distribution and measure are *mathematizing concepts* providing *mathematizing goals*.

The list of mathematizing concepts is far from being complete as yet. For example, important concepts such as covariation, sampling, and inference are still missing. The list only provides necessary, but not necessarily sufficient mathematizing concepts for reaching the mathematizing aims. For the design of a compact teaching-learning arrangement, however, again some selection needs to be made. The justification of this choice (the interconnectedness of the three identified mathematizing concepts) is provided in Chapter 3.

2.4 Aims and goals of the reflective side of mathematical literacy

After the previous section identified mathematizing aims and goals, this section likewise focuses on the reflective side of mathematical literacy. The section begins by outlining the general reflective aims (2.4.1) In analogy to the activities and concepts of the mathematizing side of mathematical literacy, the reflective goals are developed by identifying reflective activities and concepts. During this, it becomes clear that the reflective side of mathematical literacy is much less developed regarding theory than the mathematizing side. Whereas some reflective activities can be identified (2.4.2), a specification gap concerning reflective concepts is identified (2.4.3). Some proposals to close this gap are made, but ultimately this is treated as a task for empirical research. This section expands on ideas proposed by Büscher and Prediger (submitted).

2.4.1 Two reflective aims

As sketched in Section 2.1, reflection refers to the activity of critically evaluating the role that mathematics plays in the world. Section 2.3.1 already established the role of mathematics as a means and a system of communication. By further exploration of the character of mathematics as a system of communication, two general reflective aims are identified: the ability to identify the formatting power of mathematics, and the ability to evaluate expert judgements.

Identifying the formatting power of mathematics

One possible type of reflective goals is laid out by the socio-political approach of *Critical Mathematics Education* by Skovsmose (1994; 2005). His main thesis is that mathematics is not a pure and abstract force disconnected from human society (as for example argued by Hardy, 1940/2005). Rather, mathematics exercises what he calls *formatting power*: by providing the tools and concepts for investigating real-world phenomena, mathematics takes an active role in shaping society.

The formatting power is an expression of the character of mathematics as a system of communication. Although mathematics has historically been created by living people in a dynamic and open process of discovery in which the meaning of concepts gets negotiated (Lakatos, 1976), individuals find themselves before a rigid system of concepts and rules to which they have to submit if they want to engage in interpersonal communication. It is a paradox that this 'character of compulsion' (Fischer, 1988) enables communication while simultaneously hampering it: it provides the means of communication, while it *formats* the discourse by channeling it into mathematical pathways. Skovsmose refuses interpretations of the role of mathematics that paint it as abstract, gentle, and clean; instead, mathematics actively interferes with human society. It is not simply a tool for describing and understanding the world, but it remakes the world in its own image:

"[Mathematics] does not simply describe a domain; it configures this domain. Mathematics only describes as it has prescribed" (Skovsmose, 2012, p. 126).

The formatting power is exercised through mathematical modeling. Mathematical models of real-world phenomena are created in order to provide a transparent basis for making political, economic, or environmental decisions. This process of modeling, however, cannot be conceptualized as a simple translation from the real world to mathematics. Phenomena are never accessed directly but always through the system of mathematics, because the tools used for accessing phenomena are already based on mathematics and perception is influenced by mathematical concepts. In this way, phenomena come with a mathematical pre-understanding which already shapes the decisions to be made (Skovsmose, 1994).

Skovsmose stresses that this does not make mathematics inherently good or bad, but that mathematics can create 'wonders' as well as 'horrors' (Skovsmose, 2005). Mathematical modeling can enable society to notice and to deal with complex phenomena such as climate change (Barwell, 2013). It can also dehumanize the workplace by conceiving of human workers as simply a collection of factors concerning productivity of the system (Skovsmose, 2012). Mathematics can be used by governments to allocate resources in order to increase everybody's standard of living. It can also provide the language for bureaucratic systems that enable genocide (Skovsmose, 2005). Mathematics itself is neither beneficial nor antagonistic; it can simply play various roles in society.

Although mathematics thus plays an important role in society, its influence gets underestimated. Mathematics seldom is in the focus of public discourse. This can be explained as the result of a process of demathematisation (Jablonka & Gellert, 2007): once they are articulated, mathematical ideas become encoded in cultural practices and tools. These tools persist even as the ideas fade from discourse. The mathematics behind the tools becomes invisible, while still exercising its formatting power.

The character of mathematics as a system of communication leads to the formatting of society through mathematics. Formatting power gets exercised invisibly; it is built into our very system of communication and constitutes a part of what is experienced as 'reality'. It is important to know that things could be different, and that mathematics is created and applied by people. It thus becomes an aim of mathematics education for students to be able to identify the formatting power of mathematics in everyday situations.

Evaluating expert judgement

Section 2.3.1 introduced Fischer's (2001) argument that students do not need to become experts themselves, but need to be able to communicate with experts. They need to be able to describe problems in a way that enables them to consult experts. Answers produced by experts, however, are not to be taken as gospel. Experts in specific fields may not know nor care about the context in which the problem was stated. The answers of experts are formulated in terms of their field of expertise, and are thus not necessarily meaningful to the person looking for advice (Feyerabend, 1978). An informed citizen therefore needs to be able to reformulate the answers provided by experts into their own context, and judge whether the answer actually is relevant to them. This is not a task that can itself be taken over by experts, but needs to be carried out by the individual. For example, one can provide experts (or a computer) with data and ask for a test of statistical significance, and the answer will be calculated correctly. One still needs to judge whether the answer is *actually* significant in the problem context (Ziliak & McCloskey, 2009). In Fischer's (2001) words:

"One generally can rely on the technical *correctness* of the expertise, that it is up to date, and that there exists a working mutual control of the experts in a field. In questions regarding the *importance*, i.e. the held personal meaning of an expert judgement, how it is evaluated, one is reliant on one's own judgement" (Fischer, 2001, p. 152 own translation).

To illustrate, Fischer (2001) uses the picture of a judge in court to explain his views on the goal of education. The judge needs to be able to question experts to gain a clearer view on the issue at hand – but has to personally decide the correct course of action, and will be held accountable for any decision made.

This has consequences for the aims of education. Contrasting general education with expert education, Fischer distinguishes between three areas of knowledge (Fig. 2.2). *Basic knowledge*, the knowledge of concepts and simple procedures of the field; *operating*, the complex skills in problem solving and research; and *reflecting*, relating to the meaning and limitations of concepts. Basic knowledge enables the communication with experts, whereas reflective knowledge enables the judgement of expertise. Both areas of knowledge are an important goal of education. Operating, however, is of less importance – although problem solving might be a worthwhile activity for generating basic and reflective knowledge, it does not itself constitute a goal of general education. For organizing subject matter, Fischer calls for a *"reduction of aspirations regarding operating, and increased aspirations regarding reflecting"* (Fischer, 2001, p. 155, own translation).

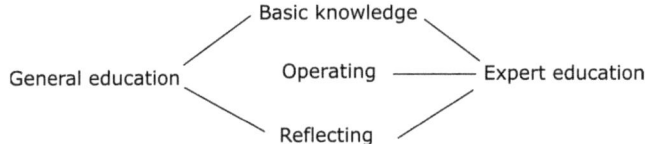

Fig. 2.2: Operating is an area of knowledge specific to expert education (translated from Fischer, 2001)

Thus, another reflective goal of mathematical literacy in statistics is *evaluating expert judgements*. Regarding statistics, the need to critically evaluate statistical expertise is as old as the discipline itself. An interesting case can be observed with the English translation of the French work of Quetelet (1842/1994). Quetelet uses his new idea of using averages to compare large groups of people to explain differences in the height of conscripts:

"The average height of conscripts [...] was 1.677 metre. On the other hand, in the ancient department of the Appenines, of which Chiavari was the chief place, the country mountainous, without industrial occupations, extremely poor, and where the men toil from a very early age and are ill fed, the average stature of the conscripts [...] was 1.560 metre" (Quetelet, 1842/1994, p. 75).

Quetelet makes a convincing case that malnourishment and bad living conditions in general can have an impact on growth. His translator, however, in a footnote on the above excerpt, arrives at a different conclusion:

"The translator is firmly persuaded that Dr. Villermé and M. Quetelet have failed to detect the real cause of difference in stature in those two departments: it is a question purely of *race*, and not of feeding or locality. The taller conscripts were Saxons, [...] the shorter conscripts [...] were descendants of the ancient Celtic population of that country" (Quetelet 1842/1994, p. 75).

The translator uses the same data as Quetelet, yet in a way that subscribes to the ideology of 19th century British colonialism. Yet both experts use essentially the same argument concerning the difference of averages; on these grounds alone, none of the two interpretations can be given precedence over the other. Citizens need to be able to evaluate such expert judgements, so that they do not unwittingly reproduce the possibly hidden ideologies of experts.

Thus, learners do not only need the basic knowledge that enables them to communicate with experts. The system of communication constructed through mathematics might require disregarding information that would have individual relevance. Answers produced by experts might in the best case simply not fit their own interests, and in the worst case reflect an expert's ideology. They need to be able to reflect on expert judgement in order to become informed citizens.

2.4.2 The reflective goal of mastering reflective activities

The reflective aims show how important it is for learners to reflect on mathematics. So far, however, it is not entirely clear what constitutes a reflection that supports the reflective aims. Mathematics education research exhibits a variety of different conceptualizations of reflection. Freudenthal (1983b), for example, describes reflection as a 'change of standpoint' regarding situations. To Freudenthal, however, reflection does not aim at the relationship between mathematics and society, but rather constitutes a preliminary step to formal proof. Such a kind of reflection would be more suited to the mathematizing side of mathematical literacy. Skovsmose (1998) in contrast provides a framework of different types of reflection focusing on the mathematical modeling process, the social impact of the mathematical model under reflection, and its relation a student's own life – an example of reflections that support the reflective aims.

The notion of reflection seems to allow for different meanings. Alrø and Skovsmose (2003) refer to reflection as an 'explosive concept', one that cannot be easily and comprehensively captured, but rather has a multitude of meanings and relations to other concepts. One thus should not attempt to 'clarify' the concept, as this is not possible without pruning it of important aspects. Skovsmose (1994) also rejects the idea of a canonical list of intended reflective knowledge:

"The expression 'reflective knowledge' indicates the existence of some degree of explicability and perhaps the existence of some sort of authorized body of knowledge, and this is deceptive. We have to look for this particular competence in terms of dispositions and abilities" (p. 101).

To emphasize his open-ended notion of reflection, Skovsmose instead opts to speak of a process of reflective *knowing*. This resonates with Fischer's conceptualization of a general competence of *reflecting* that is to be applied to all kinds of expert judgements. Both authors thus characterize reflection as an activity. This calls for an investigation into which smaller activities can serve as reflective goals for the design of a teaching-learning arrangement. Compared to the mathematizing activities, however, more theoretical work needs to be done to identify suitable reflective activities.

The works of research mathematician and philosopher Rudolf Wille (1988, 1995, 2000, in English 2008) can provide a starting point for the identification of reflective activities. In his program of *Generalistic Mathematics*, Wille aimed at 'restructuring' mathematics in a way that makes the subject principally understandable and relatable to society. This required philosophical work on mathematical concepts and methods with the following tasks (translated from Wille, 1995):

(T1)   describing reasons and consequences
(T2)   explicating aims and purposes
(T3)   uncovering meanings and interpretations
(T4)   revealing connections and relations
(T5)   identifying patterns of thought and action
(T6)   analyzing concrete circumstances for theory
(T7)   denominating risks and limits

(T8)   considering historical understandings and conditions
(T9)   uncovering inappropriate obstructions
(T10) utilizing rich means of expression, particularly the common language.

These tasks can provide the activities constituting the reflective goals. Yet again, some selection has to be made. Identifying the patterns of thought (T5) behind some part of mathematics (or a tool utilizing mathematics) can be a first step in identifying the formatting power of mathematics. By explicating aims and purposes (T2), individuals can uncover the hidden assumptions behind experts' judgements. And denominating risks and limits (T7), can be a first step to find protection from unintended or possibly harmful effects of the formatting power. As such, this study conceptualizes *explicating aims and purposes*, *identifying patterns of thought*, and *denominating risks and limits* as the reflective activities constituting the reflective goals that need to be reached for the reflective aims of mathematical literacy.

2.4.3 The specification gap of reflection

The investigation into the mathematizing side of mathematical literacy showed how activities and concepts both can play an important part in specifying the learning goals. Whereas reflection is commonly conceptualized as an activity and work needed only be done to specify smaller reflective activities as reflective goals, the case is different with a focus on concepts. Reflection seldom is explicitly conceptualized through concepts. For the mathematizing goals, the 'big ideas' or 'core concepts' of statistics describe how ways of thinking can be structured by concepts of the subject domain that are necessary to reach the mathematizing aims. The list of mathematizing concepts then allows a closer specification of the learning content and an increased focus for teaching-learning arrangements.

Such a focus also is needed for the reflective aims. There is no reason why reflection should not be conceptualized as being structured through *reflective concepts*. However, there exists a lack of 'big ideas of reflection' that could inform the identification of such reflective concepts. A *specification gap* (Büscher & Prediger, submitted) emerges: the concepts to be learned for reflection, and indeed the very notion of what a reflective concept could be, are unclear.

Some inspiration might be drawn from Abelson (1995), who structures his instructional textbook on statistics not in a traditional way along the lines of mathematizing concepts, but along the themes of rhetoric, 'fishiness', articulation, generality, interestingness, and credibility. These themes are not mathematizing concepts, but instead are concerning with *how* mathematizing concepts are utilized. Abelson is not the only author who holds concepts other than mathematizing concepts to be important; however, systematic research into what constitutes possible reflective concepts is still needed. As a first step towards more systematic research, this study now identifies some possible reflective concepts (see also Büscher & Prediger, submitted).

Manipulation

The construct of statistical literacy can provide a subject-specific source for the identification of possible reflective concepts. As illustrated above, statistical literacy deals with "people's ability to act as effective 'data consumers' in diverse life contexts" (Gal 2002, p. 3). In their daily lives, people are confronted with (statistical) models produced by others, and statistical literate individuals are able to understand the statistical information imparted. Thus, statistical literacy is part of the reflective side of mathematical literacy in statistics.

Among knowledge elements such as statistical and mathematical knowledge that make up the knowledge base for statistical literacy, Gal (2002) also emphasizes the importance of a critical stance and the critical knowledge required to act on it:

"[Individuals] have to be concerned about the validity of messages, the nature and credibility of evidence underlying the information or conclusions presented, and reflect upon possible alternative interpretations of conclusions conveyed to them" (Gal 2002, p. 17).

Gal realizes this critical knowledge to be developed by a list of 'worry questions' that individuals should have prepared for interpreting statistical information. These include questions like: where did the data come from? Could the sample reasonably lead to valid inferences about the target population? Are the given graphs drawn appropriately? Is correlation confused with causation? Is there information missing (Gal, 2002, p. 16)?

These worry questions concern the nature of the sampling used, the reliability of instruments, distortions produced by manipulative graphs, and intentionally falsified statistics by 'conveniently forgetting' statistically relevant details. Thus, they aim at uncovering faulty use of statistics that nevertheless gets presented as evidence. The worry questions here are conceptualized as belonging to the reflective concept of *manipulation*. Such manipulation does not have to be intentional, but it always concerns the faulty use of statistics. It presents a reflective concept, because individuals should learn to be sensitive to manipulation when critically evaluating statistical information.

Contextual relevance

The concept of manipulation aims at uncovering ways in which statistics is intentionally falsified, making the statistics used formally wrong: it is simply not a correct way to use inferential statistical methods when the sampling was intentionally biased. Simple wrong use of statistics, however, only makes a small part of what reflection should uncover. In a case of manipulation,

Fischer's (2001) assumption of the correctness of the expertise is not given: even formally correct advice of experts needs to be reflected regarding its importance for the seeker of advice. This cannot be done by drawing on mathematizing concepts only, as these belong to the domain of the expert. For judging expert advice, the *contextual relevance* of the advice needs to be taken into account. This could thus provide another reflective concept to be developed.

Perspectivity

Another approach is to search for concepts that would be needed to identify the formatting power of mathematics. Formatting power is exercised by mathematical models that do not simply describe phenomena, but already prescribe the ways to view the phenomenon. Interest groups can utilize mathematical models to justify their views, implying an objective description where in fact the models were chosen to create this perspective on the phenomenon. Even though this is a type of manipulation, such use of models, however, at no point is formally wrong. To identify this formatting power, individuals need to consciously look for other ways the phenomenon could have been described. This could provide insights into which aspects of the phenomenon were intended to be highlighted through the mathematical model, and which were intended to be dropped. By drawing on the reflective concept of *perspectivity*, one can find the subjective traces in only illusory objective mathematical modeling.

Closing the specification gap

This specification gap hinders the design of teaching-learning arrangements and needs to be overcome in order for research into students' reflections to become more systematic. The concepts of manipulation, contextual relevance, and perspectivity form only a proposal of possible reflective concepts; yet some first step needs to be undertaken. The specification gap, however, puts this study in a paradoxical situation: how can a teaching-learning arrangement for mathematical literacy be designed, if the reflective goals are unspecified?

Two answers can be given. First, the identification of reflective activities has already provided one reflective goal that can inform the design of a teaching-learning arrangement (i.e. by designing to initiate such reflective activities). Second, the identification of reflective concepts is treated as a task with an empirical component. Chapter 6 illustrates how this is not a theoretical task only, but how insights into students' learning processes can support the specification of the content to be learned. The specification gap needs to be closed from two sides: theoretical foundation (Chapters 2 and 4), and empirical grounding (Chapters 7 and 8). This study aims to provide contributions to both.

2.5 The concept of measure as the central learning content

This chapter has identified the mastering of mathematizing and reflective activities and the development of mathematizing and reflective concepts as goals of a teaching-learning arrangement for mathematical literacy in statistics. Regarding the mathematizing concepts, some sequencing of the concepts needs to be identified to structure students' learning trajectories.

This study chooses the mathematizing concept of measure as the central learning content. Two reasons can be given: first, the following chapter will argue that the mathematizing concept of measure can be specified as a prerequisite to the development of the mathematizing concepts of distribution and variability. It also provides connections to students' individual thinking. Second, measures can play a central role in the reflection of the role of statistics in society:

"[Measures] create descriptively and almost always normatively determined realities that one can hardly escape from. They provide us with glasses through which we observe the world around us, they facilitate and enable decisions or enforce them. At any rate, they structure many areas of our human co-existence in a specific direct or indirect way" (Lengnink & Peschek, 2001, p. 75).

Statistical measures shape society by projecting formatting power, and experts often employ statistical measures in their judgements. Reflecting on statistical measures seems a valuable and rich learning opportunity for developing reflective concepts and mastering reflective activities.

Thus, the mathematizing concept of measure promises to provide a learning content that can integrate the development of mathematizing and reflective concepts and the mastering of mathematizing and reflective activities. A closer look on this concept, however, is necessary. The following chapter explores this mathematizing concept of measure.

3 The mathematizing concept of measure

During the specification of mathematizing and reflective goals in Chapter 2, the development of the mathematizing concept of measure emerged as a mathematizing goal of mathematical literacy in statistics. It was also proposed that this concept is in need of a thorough conceptualization, so that the learning content of a corresponding teaching-learning arrangement can be specified. This chapter argues that the mathematizing concept of measure is indeed in need of a conceptualization (Section 3.1.1), and then proposes such (Section 3.1.2). The conceptualization provides a framework that allows to further specify learning goals as well as to describe the starting points of learners' reasoning (Section 3.2). Finally, the mathematizing concept of measure is identified as a necessary precursor to the development of a goal common to literature, the development of Informal Statistical Inference (Section 3.3).

3.1 A conceptualization of measure

Measures are central to statistics. However, statistics education research has mostly focused on few specific measures in isolation instead of the whole concept (Section 3.1.1). These isolated insights are adopted in a more general conceptualization of measure (Section 3.1.2).

3.1.1 Measures and the mathematizing concept of average

Although central to statistics, little attention has been given to the concept of measure itself. The concept is in need of a conceptualization that allows to specify what exactly learners have to learn to develop the concept of measure. Research in statistics education so far, however, does not focus on this concept, but touches on the concept of measure mostly through the mathematizing concept of average, which is related to the measures of mean, mode, and median.

Mokros and Russell (1995) link the mathematizing concept of average to the concept of representativeness. They identify mode, algorithm, reasonable, midpoint, and mathematical balance point as approaches used by students for interpreting the mean. They argue that an early focus on the algorithm for computing the mean actually hinders the development of an understanding based on representativeness. Konold and Pollatsek (2002) distinguish 'formal properties' such as measures of average not necessarily corresponding to an actual observation and 'interpretations' such as the average as signal in a noisy process. They argue that instruction often is focused on the former, whereas the latter actually is more important and propose the idea of 'signal and noise' as an important interpretation of average. Konold et al. (2002) propose to build on students' conceptions of typicality and their use of 'modal clumps', small intervals of high densi-

© Springer Fachmedien Wiesbaden GmbH, part of Springer Nature 2018
C. Büscher, *Mathematical Literacy on Statistical Measures*,
Dortmunder Beiträge zur Entwicklung und Erforschung des
Mathematikunterrichts 37, https://doi.org/10.1007/978-3-658-23069-2_3

ty around the mode. Since the students use modal clumps for summarizing data, they propose to use modal clumps to develop an understanding of average. Watson (2007) shows how the creation of cognitive conflict between different interpretations of average can support the development of the concept of average. Makar (2014) identifies several key concepts that can support the development of an understanding of average: reasonableness, outliers, typical as most common, and comparing groups.

Although these approaches provide some directions for students' development of the mathematizing concept of average, they do not provide much direction for a specification of concepts to be learned. They provide a wide range of elements 'supporting' the development of average, ranging from specific advice about instruction sequence (do not focus on algorithm too early), to task design (use comparing groups tasks), to specification of related concepts (reasonableness, typical), to advice regarding general theories of learning (build on students' conceptions, introduce cognitive conflict). However, they eschew the question what an average actually *is*, and what therefore needs to be learned about averages. Neither do they explicate the relationship between the mathematizing concept of average and the measures of mean, mode, and median.

A conceptualization of measure is in order that allows the specification of the learning content of a teaching-learning arrangement focusing on mathematical literacy in statistics. Such a conceptualization should not only clarify measures of center such as mean, mode, and median, but other measures such as the interquartile range as well. In the following, such a conceptualization of the mathematizing concept of measure is given.

3.1.2 Conceptualizing measure

The concept of measure seems to be in need of clarification. One way for such a clarification is the investigation of the use of measures in language (cf. Wittgenstein, 1953/2008). To uncover relevant concepts suitable for a conceptualization of measure, this thesis investigates how measures can be used in a series of different example assertions related to statistical measures. Such assertions can appear very different on a surface level: for example, the assertions "the mean is a measure of center" and "the low median household income shows the income inequality in Germany" both use measures, but in very different ways. However, this thesis introduces only a small number of concepts that allow a conceptualization comprising all these different uses (see advance organizer in Fig. 3.1).

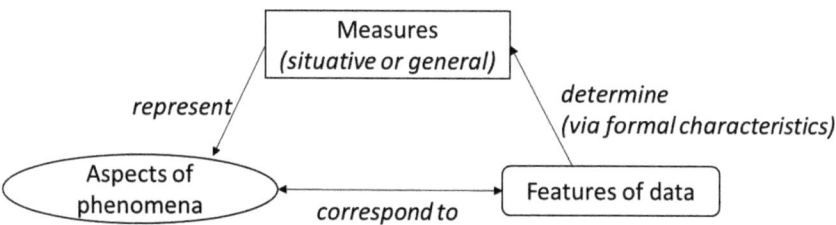

Fig. 3.1: Measures represent aspects of phenomena and are determined by features of data through their formal characteristics

Through investigating the use of measures in language, this thesis argues that measures can be situative or general measures; that measures are used to represent aspects of phenomena; that measures and aspects of phenomena correspond to features of data; and that measures are determined through their formal characteristics.

This conceptualization is be elaborated and supplied with meaning through the following investigation of assertions. The following two assertions can be commonly observed in explanations of specific measures.

(1) "The mean is a measure of center."
(2) "The median is a measure of center."

Assertions such as (1) and (2) often appear as clarifying remarks in definitions of measures. Although mean and median are two measures with completely different procedures of calculation, these assertions assert a similarity between the two different measures. Mean and median belong to the same *class of measures*, namely center. Classes of measures comprise all measures that have similar uses. Another example of measures and class are the measures of interquartile range and standard deviation belonging to the same class of measures of spread.

Assertions about measures, however, can also take another form:

(3) "The interquartile range comprises the middle 50% of the data."
(4) "The mode is the value with the most data entries."

Whereas assertions (1) and (2) referred to classes of measures assertions (3) and (4) refer to the actual data (the middle 50%, the most data entries). The value of a measure can generally be estimated by looking at the data. This is because measures correspond to some *features of data* such as the middle 50% of the data or the value with most data entries. A feature of the data can be any identifiable part of the actual data at hand. Thus, every data set shows several features of data at once. Measures *correspond* to specific features of data, which allows a visual estimation of their values for actual data: to estimate the value of the mean, one can use the feature of data of the location of the majority of data entries.

Other assertions focus on a different facet of measures:

(5) "The interquartile range is the difference between first and third quartile."
(6) "The median is resistant against outliers."

Assertions (5) and (6) make no reference to class of measure or feature of the data. Instead, they refer to the procedure of calculation for the interquartile range and the characteristic of robustness of the median against outliers. Thus, they refer to general behavior of the measures given any data. Calculation and robustness belong to the *formal characteristics* of a measure: its form, its procedure of calculation, its abstract behavior with different kinds of hypothetical data. The value of a measure given actual data then is *determined* by its formal characteristics.

Other types of assertions relate to the use of measures in argumentation:

(7) "The average household income shows that households in Germany generally have an acceptable standard of living."
(8) "The low median household income shows the income inequality in Germany."

Assertions (7) and (8) show that the class of measures does not determine their use. Although mean and median belong to the same class of center, and both assertions refer to the same body of data, assertions (7) and (8) differ in their assessment of the situation. Using the mean, one could argue that things are fine; using the median, one could argue that they are not. The measures do not differ in class of measure; in these assertions, they differ in the *aspect of the phenomenon* each measure *represents*. An aspect can be any identifiable part of a larger phenomenon. For example, the larger phenomenon of household income in Germany comprises the different aspects of the phenomenon of standard of living and income equality. Whereas the mean represents the general standard of living, the median represents the balance between poor and rich households. Both of these aspects belong to the same phenomenon of household income in Germany.

The same measure, however, can also be used in different assertions:

(9) "The high average grade of my students shows the high quality of my instruction."

(10)"The high average grade of your students shows the strength of this year's cohort."

The aspect of the phenomenon that is represented by a measure, however, also is not predetermined by its class or formal characteristics. Instead, assertions (9) and (10) show that the same measure can be used to represent different aspects of phenomena. In this hypothetical scenario, a teacher could use the mean to represent the aspect of quality of instruction for the phenomenon of her own students' schooling (9). For students from other classes, the same measure could represent the general strength of the cohort (10). Thus, the aspects of a phenomenon represented by a measure neither depend on the class of measure nor on the specific measure itself. A single measure can represent a variety of different aspects of different phenomena. Using a measure as evidence in argumentation thus means to choose a measure to represent a specific aspect of the phenomenon.

Still other types of assertions use other types of measures:

(11)"The high gross domestic product shows the economic strength of Germany."
(12)"The rising adult literacy rate shows the improvement of the educational system in Burkina Faso."

All measures employed so far are independent of the actual phenomenon: median and mean can always be used where there is the right form of data. Assertions (11) and (12) show other types of measures. The gross domestic product represents the aspect of economic strength for the phenomenon of Germany's larger economy; the adult literacy rate represents the aspect of the outcome of education for the larger phenomenon of Burkina Faso's educational system. In contrast to mean and median, however, these two measures are bound to their specific phenomenon. Whereas the mean can be used for many different phenomena, the gross domestic product can be used only for economic systems. Gross domestic product and adult literacy rate are *situative* measures that have their specific roles for specific phenomena, whereas mean and median are *general* measures that can be employed for a variety of phenomena.

Defining measure

The constructs of aspects of phenomena, features of data, and formal characteristics now allow this thesis to define statistical measures as *data-based formalized representations of aspects of phenomena*. They can be characterized through their represented aspects of phenomena, corresponding features of data, and determining formal characteristics. A *class* of measures are all measures representing 'similar' aspects of phenomena – an intentionally vague definition,

as the similarity between mean, median, and mode for example is hard to articulate without already pointing to them all being 'measures of center'. Tab. 3.1
shows the characterizations of some general measures. No entry itself, however,
presents a full characterization, as a complete listing of all formal characteristics, features of data, and possible aspects of the phenomenon is possible.

Tab. 3.1: Measures are characterized through their class, formal characteristics, corresponding
features of data, and represented aspects of the phenomena

Measure	Class of measure	Formal character-istics	Features of data	Aspects of the phenomenon
Mean	Center	Calculation through arithmetic mean; sensitive to outliers	'Hill' of unimodal distributions	Estimated expectation
Median	Center	Calculation by counting; insensitive to outliers	Boundary for equal halves	Balance be-tween two equal groups
Interquartile range	Spread	Calculation through median; Middle 50% of data	Dense area in middle	Expected devia-tion from ex-pectation

Tab. 3.1 necessarily stays general, as without an example phenomenon, the
represented aspects of phenomena need to stay abstract. Figure 3.2 gives an
example of the relevant constructs concerning measures so far, set in the example phenomenon of Arctic sea ice. The phenomenon of Arctic sea ice generates a
wealth of data that scientists can use in their investigations. The whole phenomenon consists of many different aspects of the phenomenon, and scientists use
different measures to *represent* these aspects. For example, the situative measure of daily ice growth is used to represent the melting process of the ice, allowing for insights into the volatility of changes. This aspect of Arctic sea ice *corresponds* to a feature of the data, the variation observed in the data. Another situative measure, the monthly average extent, is used to represent the general state
of the ice, allowing for long-term insights into the system as a whole. This aspect corresponds to the feature of high density areas in the data. The measures'
formal characteristics then *determine* the actual values of the measures based on
the features of data.

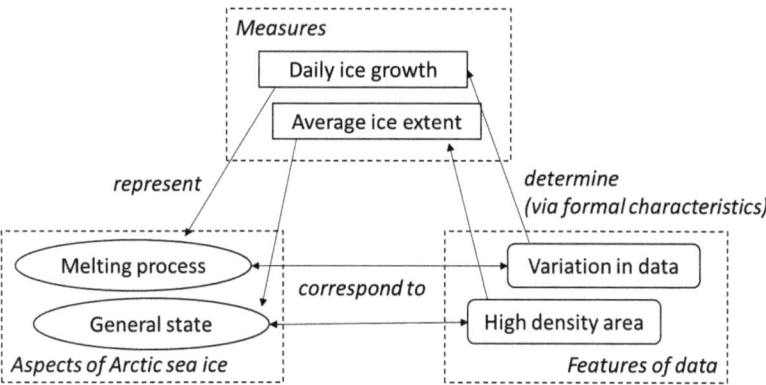

Fig. 3.2: Situative measures of the phenomenon of Arctic sea ice and their represented aspects of the phenomenon and corresponding features of data

These considerations, however, refer to the normative use of measures by experts. The following section turns towards learners' use of measures in order to identify starting point for their development of measures.

3.2 Learners' intuitive situative measures

General measures and situative measures like the gross domestic product or the average ice extent both are expressions of expert reasoning. Learners' reasoning, however, does not correspond directly to such formalized notions, but is based on intuition (Fischbein, 1999). This means that learners use *intuitive* situative measures that do not necessarily correspond to experts' situative or general measures.

The concept of measure was proposed as a concept that allows to connect to learners intuitive reasoning for developing mathematizing concepts. A great deal of empirical studies have already investigated learners' intuitive reasoning about statistics. The conceptualization of measure proposed above can serve as a conceptual framework for describing these intuitive starting points. Although these studies originally do not frame learners' thinking in terms of their use of measures, they still can provide some insights into learners' use of measures.

Makar & Confrey (2005) find that prospective teachers use nonstandard language when talking about variation to integrate statistical concepts that would formally belong to distinct statistical concepts. Instead of using measures that would neatly conform to the classes of center or spread, the prospective teachers in the study used terms like 'clustered' or 'spread out' to represent aspects of phenomena, relating to classes of center, spread, and shape of distributions simultaneously. In a study by Makar & Confrey (2003), prospective teachers also

identified 'clumps' and partitioned data of students' achievement after an en-
richment program into 'chunks' to represent the aspects of the phenomenon of
groups that profited from the program and groups that did not. When comparing
distributions, the students studied by Konold et al. (2002) focused on 'modal
clumps', which are ranges around the mode in the center of data, to represent
the aspect of phenomenon of the amount of roadkill on a usual day. In the study
by Bakker & Gravemeijer (2004), students engaged the phenomenon of battery
life by representing aspects of the phenomenon like reliability and chance.
Konold, Higgins, Russell, & Khalil (2015) show how learners use different
perspectives on data, ranging from using single data entries as pointers to specific events (i.e. identifying aspects of the phenomenon concerning single
events) to perceiving the data as an aggregate with emergent properties of its
own (i.e. identifying larger aspects of the phenomenon). They argue that although the aggregate perspective is an important goal of instruction, learners
should be able to flexibly change between perspectives, i.e. they should learn to
flexibly use different situative and general measures.

All these studies show examples of learners using situative measures to represent aspects of the phenomena that do not neatly fit to general measures or
classes of measures. They also show a wide range of features of data identified
by the learners, such as modal clumps or different chunks of data. Konold et al.
(2002) call for using these learners' situative measures as a starting point for
learning processes by engaging in the mathematizing activity formalizing:

"For instruction to build on the idea of modal clumps, students will need to begin to formalize and
study them. The need for formalization could be introduced by having students explore communication problems that would result from each student using his or her own ad hoc methods for composing them" (p. 6).

Similar calls have been voiced by the other authors. As such, research has
acknowledged the potential of learners' situative measures for the development
of mathematizing concepts. However, there seems to be an overemphasis on
features of data: studies such as those by Konold et. al (2002) or Makar & Confrey (2003) seem to imply that learners generally tend to focus on features of
data such as modal clumps or chunks. Using an analytical framework for the
investigation of learning processes that exclusively focuses on the features of
data identified by students, however, would omit possible insights into how
learners structure phenomena. The conceptualization of measure above allows
to also take into account how students structure phenomena through aspects of
the phenomenon in order to fully understand their learning processes.

Research into learners' situative measures thus has mostly focused on features of data. More insight is needed into the aspects of phenomena represented
by learners' measures and into the development of general measures from learners' intuitive situative measures. This thesis provides contributions to that aim

by illustrating a hypothetical learning trajectory (Chapter 4) and by empirical analyses of learners' development of measures (Chapters 7 and 8).

This thesis has placed the concept of measure as a central mathematizing goal of mathematical literacy in statistics. However, large parts of statistics education research instead hold the construct of Informal Statistical Inference (ISI) to be central. In the next section, this thesis illustrates the connections between measures and ISI.

3.3 Measures and Informal Statistical Inference

In recent years, increased attention has been given to the constructs of Informal Statistical Inference (Makar & Rubin, 2009; 2018) and Informal Inferential Reasoning (IIR, see Zieffler, Garfield, delMas, & Reading, 2008; Makar et al., 2011) in statistics education research (Ben-Zvi, Bakker, & Makar, 2015). The introduction of ISI and IIR marked a significant shift in statistics education research, akin to the introduction of the constructs of exploratory data analysis (Tukey, 1977) or statistical thinking (Wild & Pfannkuch, 1999).

Makar and Rubin (2009) initiate this shift through normative assertions about the general aims of statistics education:

"It is vital that the focus in using statistical tools is embedded in the reason that we do statistics – to understand underlying phenomena" (Makar & Rubin, 2009, p. 84).

"Focusing on statistical phenomena entails understanding the statistical investigation cycle *as a process of making inferences*. That is, it is not the data in front of us that is of greatest interest, but the more general characteristics and processes that created the data" (Makar & Rubin, 2009, p. 84).

In professional statistics practice, these aims belong to the domain of statistical inference. Statistical inference is a formally advanced concept with its theoretical foundations in probability theory, making its introduction into school practice unpractical. To enable the general idea of statistical inference for school practice, Makar and Rubin (2009) introduce the construct of *informal* statistical inference by identifying three key principles of statistical inference: (1) generalizing beyond the given data, e.g. by predicting instead of describing data; (2) using data as evidence for these generalizations; and (3) using probabilistic language. Assertions that fulfill these three key principles yet do not make use of formal statistics and probability calculus are *informal* statistical inferences. The twin constructs of ISI and IIR have stimulated statistics education research (see Makar & Rubin, 2018), and students' inferential reasoning has been investigating from kindergarten and early education (e.g. Makar, 2016) to primary school (e.g. Paparistodemou & Meletiou-Mavrotheris, 2008; Makar, 2014), secondary school (e.g. Pfannkuch, Arnold, & Wild, 2015), up to college and university level (e.g. Pfannkuch, Budgett, & Arnold, 2015).

Doubts, however, can be raised regarding this normative orientation on statistical inference as the central learning content (White & Gorard, 2017), and this thesis has also eschewed the construct of ISI in the identification of learning aims and goals. The reason is that this thesis holds ISI to be yet lacking a convincing normative foundation.

Parts of ISI conform to the normative framework introduced in Chapter 2. ISI emphasizes the utility of statistics in investigating phenomena. This is done by generalizing characteristics of phenomena and predicting their behavior in the future or after some given treatment. This can be seen as belonging to the mathematizing aim of investigating hypothetical situations. As such, learning ISI can support at least one mathematizing aim.

Other parts do not conform to the normative framework. ISI and IIR are frequently justified by their relation to statistical practice. ISI can be seen as a precursor to expert reasoning (e.g. Garfield, Le, Zieffler, & Ben-Zvi, 2015). The very notion of *informal* statistical inference emphasizes the idea of letting students engage in experts' practice without experts' concepts. Yet Chapter 2 illustrated how the development of expert reasoning can be no goal of general education. Fischer (2001) emphasizes that students do not need to become experts themselves, but that they need to be able to communicate with experts. And more importantly, in order to become informed citizens, they need to be able to reflect on and evaluate experts' judgements. An uncritical assimilation of expert practice (i.e. statistical inference) runs contrary to the reflective aims of mathematical literacy.

This should not mean that ISI does not provide value to statistics education, as indeed the use of statistics to investigate phenomena is important to understand the role of statistics in the world. Fischer (2001) also holds practice to be important – if it is just not an end in itself, but only the means for learning to communicate with experts and reflect their judgements. Under this perspective, ISI still holds value – as a means to the end of mathematical literacy.

The conceptualization of measure proposed here, however, also follows the same purposes that have led to the conceptualization of ISI. Makar and Rubin (2009) reject approaches to teaching that reduce statistics to mere description of data through simple charts and the simple computation of averages by emphasizing the importance of the phenomenon to be investigated. The conceptualization of measure proposed here follows the same line of thought: Without giving attention to the phenomenon, i.e. without explicitly focusing on the aspects of a phenomenon represented by a measure, the use of this measure remains incomplete. Instruction that uses measures only for computation overemphasizes formal characteristics of measures and features of data; it ignores the aspects of phenomena. A focus on measures as proposed here thus only drops the inferential practice from an ISI approach; instead of focusing on 'probabilistic generalizations from data' (Makar & Rubin, 2009), it favors non-probabilistic generalizations from data. Finally, a key principle of ISI is the use of data as evidence, and measures fulfill this role of evidence. Thus, learning to use measures seems a prerequisite for the development of ISI.

It again needs to be emphasized that it is not the idea of ISI, but its normative foundations that are problematized here. In light of the nature of a compact teaching-learning arrangement, the mathematizing concept of sampling has been dropped from the mathematizing goals, because the concept of measure has been deemed more urgent to develop. Given longer possible learning trajectories, sampling should again be incorporated, as it is an integral part of statistics and also needed to evaluate experts' judgments. The mathematizing concept of measure, however, is an epistemological predecessor to the development of more advanced mathematizing concepts such as distribution.

With this conceptualization of measure, the question remains how learners can develop this concept. To provide answers to this question, the next Chapter draws on epistemological theories to provide a hypothetical learning trajectory from learners' situative measures to general measures.

4 Developing concepts and mastering activities

Chapter 2 proposed the mathematizing concept of measure as a first concept to be developed in a longer learning trajectory towards the mathematizing goals of mathematical literacy in statistics. Chapter 3 introduced a conceptualization of measure that allows to describe general measures to be developed as well as learners' situative measures. This chapter introduces the learning-theoretical background of this thesis (Section 4.1). This background explicates the connections between the goals of developing concepts and mastering activities (Section 4.1.1), as well as the importance of context for concept development (Section 4.1.2). The learning-theoretical background then allows to describe a hypothetical learning trajectory from learners' situative measures towards the mathematizing goals (Section 4.2.1). Regarding the reflective side of mathematical literacy however, a realization gap becomes apparent, as the description of a hypothetical learning trajectory towards the reflective goals of mathematical literacy remains impossible for the current state of theory (Section 4.2.2).

4.1 Learning-theoretical background

To provide a learning-theoretical foundation, this thesis draws on and relates to two different theoretical frameworks: The Theory of Conceptual Fields (Vergnaud, 1996) and the theory of Situated Abstraction (Noss & Hoyles, 1992). These frameworks establish a link between activities and concepts and the importance of context. This can then inform the design of a teaching-learning arrangement. The two frameworks also allow capturing the learning processes at a micro level (see Chapter 5).

4.1.1 The relationship between concepts and activities

Chapter 2 proposed mathematizing and reflective activities and concepts as separate goals in mathematical literacy. From a learning-theoretical point of view however, activities and concepts can be conceptualized as closely linked. One theory that links these two constructs is the Theory of Conceptual Fields developed by Vergnaud (1996; 1998; 2009). Vergnaud (1996) develops a theory of learning to accomplish a complex task:

"Finally a theory of learning and cognitive development must offer at the same time a way to understand how knowledge is developed in situations, owing to the subject's activity and to other subject's cooperation, and how it can ultimately take the shape of natural language texts, and of highly formalized representations" (Vergnaud, 1996, p. 238),

The Theory of Conceptual Fields is his answer to this task. In this theory, an individual's knowledge is strongly connected to the situations this individual encountered and the actions undertaken in these situations, thus strongly empha-

© Springer Fachmedien Wiesbaden GmbH, part of Springer Nature 2018
C. Büscher, *Mathematical Literacy on Statistical Measures*,
Dortmunder Beiträge zur Entwicklung und Erforschung des
Mathematikunterrichts 37, https://doi.org/10.1007/978-3-658-23069-2_4

sizing the situativity of knowledge (Greeno, 1998). Developing ideas introduced by Piaget and Vygotsky, Vergnaud sees the role of knowledge in the organization of behavior in situations. As a first step, knowledge influences the cognition of situations, which leads Vergnaud (1996) to a central point:

"cognition is first of all conceptualization, and conceptualization is specific to the domain of phenomena to be dealt with – more precisely, to the domain of situations to be mastered" (Vergnaud, 1996, p. 224).

Two important conclusions regarding the form of knowledge can be derived from this assertion: First, since cognition is the root of all knowledge, knowledge takes the form of concepts. In this perspective, Vergnaud specifies knowledge to be learned in terms of concepts, and pleads for uncovering learners' concepts when investigating their learning processes. Second, knowledge is situated, so the research should search for the roots of knowledge in situations, and should refrain from trying to teach 'abstract' knowledge.

Although knowledge takes the form of concepts, learners' activities play an indispensable part of their learning process. Vergnaud's main assumption is that learners' activities can be implicitly guided by unconscious concepts and theorems that are tied to the situation in which those actions take place. These implicit concepts and theorems are called *concepts-in-action* and *theorems-in-action*:

"*Concepts-in-action* are categories (objects, properties, relationships, transformations, processes, etc.) that enable the subject to cut the real world into distinct elements and aspects, and pick up the most adequate selection of information according to the situation and scheme involved" (Vergnaud, 1996, p. 225).

"A *theorem-in-action* is a proposition that is held to be true by the individual subject for a certain range of the situation variables" (Vergnaud, 1996, p. 225).

Concepts-in-action are employed in theorems-in-action to structure the situation and to organize action. These constructs thus form a dialectical relationship: concepts-in-action provide the content to theorems-in-action, which in turn give meaning to the concepts-in-action. Learning consists in the individual's construction of *conceptual fields*, which comprise a set of situations to be mastered through activity along with the set of concepts the individual used to master these situations.

The suffix *-in-action* emphasizes the epistemological link between concepts and activity. In their activity, learners draw (often implicitly) on theorems- and concepts-in-action, and their concepts-in-action in turn shape their perception and activity. Learning consists of engaging in activity in order to develop concepts-in-action. Under this epistemological perspective, the mathematizing and reflective goals of developing concepts and mastering activity do not appear separate, but intricately linked. Concepts are developed by mastering activity, and activities as mastered by developing concepts.

This link between concepts and activity now allows to highlight the relationship between this thesis' central theoretical constructs: *mathematizing activities* are the activities of structuring, representing, and formalizing. *Mathematizing concepts* are the concepts-in-action organizing the mathematizing activities. Statistical measures are one kind of mathematizing concepts of particular importance for the field of study, elementary statistics; they can be used to structure phenomena, to represent aspects of phenomena, and can be the object of formalizing. The mathematizing concepts of measure, distribution, and variability serve as the intended end points of mathematizing concept development.

Chapter 2 also introduced the idea of reflective concepts. At the time, the justification for this construct relied merely on analogy to the mathematizing side of mathematical literacy. The Theory of Conceptual Fields finally can be used to provide a theoretical foundation to this construct. Reflection is an action, conceptualized here through the *reflective activities* of identifying patterns of thought, explicating aims and purposes, and denominating risks and limits. It follows that some concepts-in-action and theorems-in-action need to become active during reflective activity. *Reflective concepts* are the concepts-in-action of the reflective activities. These reflective concepts cannot be the same as the mathematizing concepts, because the mathematizing and reflective activities are not the same. The concepts of manipulation, contextual relevance, and perspectivity are examples for reflective concepts, as they can organize the explication of purposes and the denomination of risks.

Vergnaud's theory also allows to describe learners' individual intuitive conceptions as well as formal concepts to be developed within the same language of concepts-in-action and theorems-in-action. Both, individual conceptions as well as intended formal concepts can be concepts-in-action. The difference is that individual concepts-in-action are much more reliant on specific situations, whereas formal concepts-in-action can organize action in a wide range of different situations. This distinction was already introduced through the distinction between situative and general measures, which can now be substantiated with a learning-theoretical backing. *Situative mathematizing concepts* are mathematizing concepts that are bound to specific situations. Examples include situative measures, as well as aspects of phenomena that are specific to the situation at hand. The construct of situative mathematizing concepts allows to describe learners' individual thinking processes. *General mathematizing concepts* are mathematizing concepts that hold across a variety of different situations, i.e. can organize activity for whole conceptual fields. Examples include general measures such as the median as well as the larger mathematizing concepts of measure or distribution. This construct allows to describe the formal goals of concept development.

There are no reasons to proceed otherwise with reflective concepts: *situative reflective concepts* are reflective concepts bound to situations, *general reflective concepts* hold across different situations. Yet the notion of a situative reflective concept, a concept-in-action that allows to reflect on the relationship between mathematics and world, but in a situative way, nevertheless appears strange. Whereas research on learning processes has generated a wealth of terminology that allows to describe learners' individual conceptions and ways of thinking for the mathematizing side, this conceptualization appears to be unfamiliar regarding reflection in mathematics education research. This thesis explains this phenomenon of missing terminology for describing learners' precursor conceptions to reflective concepts as a special instance of the specification gap: the construct of reflective concepts in general already is in need of clarification; the situative precursors to these unspecified general reflective concepts even more so. Nevertheless, epistemological theories like the Theory of Conceptual Fields provide sound theoretical reasons for assuming the existence of situative reflective concepts; it is an empirical task to provide an existence proof and to clarify this construct.

4.1.2 The importance of context

The Theory of Conceptual Fields allows to describe learners' situative concepts and the intended general concepts using the same language, and shows the links between concepts and activities. It does, however, only sketch the process of development from situative to general concepts. To further illuminate this process, this thesis draws on an additional epistemological framework based on the construct of *situated abstractions* (Noss & Hoyles, 1992; Noss, Hoyles, & Pozzi, 2002; Pratt & Noss, 2002).

Noss and Hoyles introduce the construct of situated abstraction in order to deal with an apparent paradox: following cognitive-psychological theories of situated cognition, all knowledge is bound to specific situations (e.g. Brown, Collins, & Duguid, 1989; Greeno, 1998). Yet mathematical abstractions appear to be independent entities divorced from reality. Noss and Hoyles (1992) solve this paradox by reconsidering the notion of abstraction:

"We propose the term *situated abstraction* to attempt to capture the ways in which people make mathematical sense of everyday activities. These are sense-making devices that are *situated* in that they are derived from experiences within specific mathematizable situations. They are *abstractions* in that they are not isomorphic with these experiences: That is, they are generalized abstractions of already mathematized situations" (Noss & Hoyles, 1992, p. 448).

"[Situated abstractions] are abstractions of structure articulated within the medium at hand" (Noss, Healy, & Hoyles, 1996, p. 225).

The construct of situated abstraction thus allows to describe the early stages of learners' concept development. Formal abstractions are developed through the process of abstracting. This process, however, is not to be understood as a process of 'cutting away' the situational parts of knowledge:

"Our substantive point concerns meaning. From this perspective, abstracting – considered as a process – can be seen as a way of layering meanings on each other, rather than as a way of replacing one kind of meaning (concrete, referential) with another (abstract, decontextualized)" (Noss, Healy, & Hoyles, 1996, p. 226).

Situated abstractions thus are the opposite of meaningless: they are a collection of meaningful reference to the structure of specific situations. Yet the framework still needs to be able to describe the emergence of formal abstractions from situated ones. Pratt and Noss (2002) expand the construct of situated abstraction by emphasizing the role of context by introducing the idea of the *contextual neighborhood* of a situated abstraction:

"Schematically we think of a situated abstraction as surrounded by a *contextual neighborhood* that describes the essential conditions, purposes, and features under which the situated abstraction was constructed. Recognition of the characteristics of the contextual neighborhood by the individual assists the identification of similar conditions under which the situated abstraction is triggered" (Pratt & Noss, 2002, p. 484).

In this perspective, the main difference between situated and formal abstractions concerns the size of the contextual neighborhood, and abstraction is hence conceptualized as a gradual distinction on a continuum. Learners structure phenomena by identifying certain aspects of the situation, i.e. creating situated abstractions. Formal abstractions emerge by expanding this contextual neighborhood. They only appear to be 'decontextualized', i.e. stripped away of any connection to real situations. Under an epistemological perspective however, the emergence of formal from situated abstractions is not such a process of decontextualization:

"Decontextualization is a post hoc perspective on mathematical knowledge that does not necessarily assist in understanding the trajectories of learning, which, according to our model, emerges out of a broadening of, rather than a cutting away from, context" (Pratt & Noss, 2002, p. 487).

One supporting factor of expanding contextual neighborhoods is learners' identification of the *utility* of mathematical ideas (Ainley, Pratt, & Hansen, 2006). Learners can construct meaning for mathematical concepts by perceiving of them as useful for certain situations. The utility of mathematical concepts then can aid the emergence of formal abstractions, as new contexts can enhance or restrict the utility of a mathematical concept, thus expanding its contextual neighborhood (Pratt & Noss, 2010).

The constructs of situated abstraction, contextual neighborhood, and utility of mathematical ideas allow to describe the process of developing general mathematizing concepts. Situative measures are the situated abstractions employed by learners. They reflect the structure observed in the specific situation, i.e. the aspects of the phenomenon and the features of the data, and they are articulated within the situation. This creates utility for the situative measures. By encountering other phenomena and situative measures, learners can identify the essential conditions of the context that influenced the utility of a specific situative measure. Some situative measures might have similar utility, and some situative measures might have utility for other phenomena. This triggers an expansion of the contextual neighborhoods of the situative measures. As already emphasized, the difference between situative and general measures is that the former has utility for a large variety of situations. Mean and median have an extremely broad contextual neighborhood; they are formal abstractions. The construct of situated abstraction suggests that they might have roots in the situated abstractions of situative measures. This also conforms to the nature of concept development according to the Theory of Conceptual Fields: actions within situations form the roots of concept development. From this perspective, situated abstractions comprise concepts-in-action and theorems-in-action for specific situations, whereas formal abstractions comprise the concepts that organize behavior for whole conceptual fields.

4.1.3 Defining this thesis' central theoretical constructs

Until this point, the thesis' central theoretical constructs remained mostly implicit. This thesis needs to define the different types of activities as well as different types of concepts. The epistemological theories outlined above now can provide the elements of such definitions. The Theory of Conceptual Fields illustrates the importance of *situations*, in which learners need to engage in specific *activities* organized through specific *concepts-in-action*. From the theory of Situated Abstraction, this thesis adopts the constructs of *contextual neighborhood*, which is expanded by learners when identifying the *utility* of concepts. Formal mathematical knowledge is created by *decontextualization* of situated abstractions.

For the mathematizing side, the mathematizing aims of mathematical literacy in statistics (see Section 2.3.1) suggest two *mathematizing situations* in which individuals need to engage in activity drawing on statistical concepts: *predicting phenomena* (i.e. inferring about their behavior using statistical measures) and *ensuring communication* (i.e. using measures to share understandings about phenomena). *Mathematizing concepts* and *mathematizing activities* are those concepts-in-action and activities needed to achieve success in mathematizing situations. For example, the mathematizing concept of center can be used in creating predictions for phenomena, and measures of center provide the objective means of communication needed to ensure communication.

For the development of the mathematizing concept of measure, this thesis has identified some types of mathematizing concepts as being of special importance (Section 3.1). *Aspects of phenomena* are concepts-in-action used by learners to create identifiable structure within holistic phenomena. *Features of data* are concepts-in-action used by learners to isolate and select distinct parts of whole data sets. *Formal characteristics* are mathematizing concepts of objective form, such as a specific type of number or a method of calculation. *Measures* are data-based formalized representations of aspects of phenomena. They are the organizing concept connecting different formal characteristics, features of data, and aspects of phenomena.

This thesis has identified three mathematizing activities of special importance in mathematizing situations. With regards to measures, *structuring* is the activity of identifying new aspects of phenomena, of identifying new features of data, and of relating specific aspects of phenomena and features of data. *Representing* is the activity of adding aspects of phenomena to a measure's contextual neighborhood. *Formalizing* is the activity of decontextualizing a measure by identifying a formal characteristic. Structuring and representing are especially important in the mathematizing activity of predicting phenomena, whereas representing and formalizing are important for ensuring communication.

For the reflective side, the reflective aims of mathematical literacy in statistics (see Section 2.4.1) suggest two *reflective situations*. Learners need to achieve success in the situations of *advocating measures* and *rejecting measures* in order to evaluate experts' judgements which utilizes statistical measures. *Reflective concepts* and *reflective activities* are those concepts-in-action and activities needed to achieve success in reflective situations. For example, the reflective concept of perspectivity can be used to reject an only formally correct measure, whereas the mathematizing concept of center would not be applicable in such a reflective situation.

This thesis has identified three reflective activities of special importance in reflective situations. *Identifying patterns of thought* is the activity of identifying a new reflective concept-in-action and relating it to a specific measure. *Explicating aims and purposes* is the activity of increasing or justifying a measures' utility by drawing on a reflective concept-in-action. *Denominating risks and limits* is the activity of restricting a measures' utility by drawing on a reflective concept-in-action.

The specification gap of reflective concepts hinders any further attempts at more closely characterizing different types of reflective concepts, as so far, only three reflective concepts could have been identified (see Section 2.4.3). Nevertheless, these constructs now allow a conceptualization of how learners can attain the learning goals of mathematical literacy in statistics, provided in the next section.

4.2 The development of the mathematizing concept of measure

In Chapter 2, the mathematizing concept of measure was proposed as a mathematizing goal of mathematical literacy in statistics. Chapter 3 provided a conceptualization of measure and illustrated learners' intuitive use of measures, which raised the possibility to develop general mathematizing concepts from learners' situative mathematizing concepts This section uses the learning-theoretical foundation provided by Section 4.1 to sketch a learning trajectory towards the mathematizing goals of mathematical literacy in statistics.

4.2.1 A general learning trajectory towards the mathematizing goals

Learning trajectories are an important part of theory, as they provide holistic frameworks combining intended learning goals with theoretical foundations and empirical knowledge, enabling justified practical decision-making about learning processes (Arnold et. al, 2018). However, they can be challenging to design. Some inspiration for designing such a learning trajectory can be drawn from the *emergent modeling* heuristic of RME theory (Gravemeijer, 1999; van den Heuvel-Panhuizen, 2003; for the case of measures also see Büscher & Schnell, 2017). A central part of RME theory is the assumption that formal mathematics can emerge from students' informal activity (van den Heuvel-Panhuizen, 2001). This emergence is conceptualized through four different *levels of activity* (Gravemeijer, 1999). Learners begin with *activity in the task setting*. Their reasoning is heavily influenced by the phenomenon under investigation, and their solutions are framed within their understanding of the phenomenon. From this level, they progress to the *referential* level of activity. They start to use mathematical representations of aspects of phenomena instead of reasoning within the phenomenon itself. The third level consists of *general* activity, in which the representations start to develop a life of their own and become independent from the specific phenomena that gave rise to them. Finally, on the *formal* level the representations cease to be required for reasoning, as learners have developed the formal mathematizing concepts.

Thus, although not explicitly stated by its founders, the level principle from the RME approach conforms to the framework based on the Theory of Conceptual Fields and Situated Abstraction. This thesis takes this level principle, i.e. the principle of progressive development of formal ideas from informal activity, but re-frames it through the constructs developed earlier. Instead of changing the levels of activity, learners engage in the same three mathematizing activities; it is their mathematizing concepts that change during their learning processes. This change, however, cannot be conceptualized as a linear change from situative to general, as which the level principle might be misinterpreted. The nature of measures as situated abstractions suggests that situative and general mathematizing concepts related to measures can and do appear alongside each other.

A general learning trajectory can now be formulated (sketched in Fig. 4.1). This learning trajectory assumes that learners begin by investigating a mathematizing situation consisting of a phenomenon and some data quantifying that phenomenon.

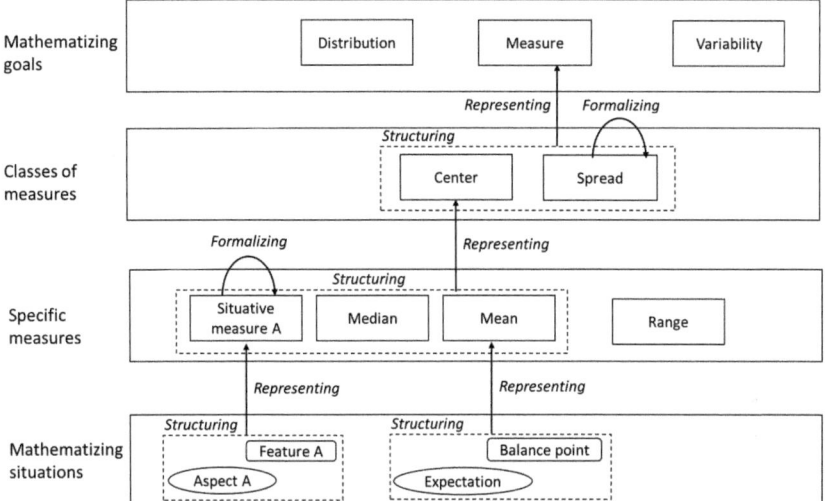

Fig. 4.1: Starting on the level of mathematizing situations, the mathematizing concept of measure can be developed through the mathematizing activities of structuring, representing, and formalizing on increasingly abstract levels of activity

In this learning trajectory, learners initially engage in the mathematizing activity of structuring to identify aspects of the phenomenon and features of the data within a mathematizing situation. Through the mathematizing activity of representing they find situative measures to represent these aspects of phenomena and correspond to the features of the data. By engaging in the mathematizing activity of formalizing, their formal characteristics are determined. Situative and general measures both form situated abstractions. General measures can also be used to represent specific aspects of phenomena. The use of situative and general measures becomes consolidated and their contextual neighborhood expanded by structuring a variety of different phenomena.

When the general measures are sufficiently developed, mathematizing activities can now be directed towards the mathematizing concepts of the measures themselves instead of only aspects of phenomena and features of the data. Structuring now is accomplished by finding similarities and differences between the different measures. Representing these specific measures then leads to the development of classes of measures. If these classes again are structured and represented, this leads to the mathematizing goal of the development of the mathematizing concept of measure. This means that the concept of measure is an organizing structure comprising all classes of measures, the classes' encompassed situative and general measures, and the specific measures' aspects of phenomena and features of the data.

This trajectory thus utilizes the constructs of mathematizing activities and concepts to show a possible connection from learners' intuitive reasoning (structuring of the situation through knowledge of the phenomenon) towards the mathematizing goals (development of the mathematizing concept of measure). However, it can only serve as a sketch of a general trajectory. Although sound theoretical reasons for the relation between situative and general measures for the development of mathematizing concepts exist, more content-specific work needs to be done. A detailed account of how the general measure of mean can be developed through which situative measures in which situations specifically is still missing. The empirical part of this thesis contributes to this gap of research.

Also missing in the hypothetical learning trajectory are pathways towards other general measures like the range, towards other classes of measures, and the other concepts of the mathematizing goals. The learning trajectory needs to be expanded to show the links between the class of measures of spread and the mathematizing goal of the concept of variability, as well as the relation between different classes of measures and the mathematizing goal of the concept of distribution. Thus, additional empirical work concerning the implementation of this hypothetical work as well as theoretical work outlining trajectories to other concepts still needs to be carried out.

For this thesis, the trajectory only serves as an outlook. As the focus lies on a compact teaching-learning arrangement, this thesis does not intend for students to develop the whole mathematizing concept of measure. Nevertheless, the hypothetical learning trajectory serves as an orienting framework. It does illustrate the vital role that mathematizing activities play in the development of mathematizing concepts: learning processes need to begin with structuring situations and developing specific situative and general measures. This thesis aims to illuminate the design of a teaching-learning arrangement focusing on just these starting points of learning.

4.2.2 The realization gap of reflection

The missing specification of general and situative reflective concepts render the formulation of a hypothetical learning trajectory for the reflective side of mathematical literacy impossible. Without start and end points, no connecting path can be conceived; and the lack of a hypothetical learning trajectory severely hinders the design of a teaching-learning arrangement that should initiate learning pathways along this trajectory.

Some empirical work concerning reflection has been carried out however. Skovsmose (1998) describes a project course in a Danish school in which students used mathematics to investigate problems surrounding actual life in the community of their small village. Based on observations made during the project, he proposes a framework of four different types of reflection. *Mathematics-oriented reflection* refers to questions regarding the correctness of calculations;

model-oriented reflection refers to questions regarding the relationship between context and mathematical model; *context-oriented reflection* refers to the impact mathematical solutions can have on the actual context; and *lifeworld-oriented reflection* refers to personal meaning of mathematics for individual students. Skovsmose, however, uses these types of reflections to clarify the notion of reflection instead of proposing a theory of how to initiate reflection in the classroom.

Some other studies do focus on reflection in the classroom. Peschek, Prediger, and Schneider (2008) emphasize the need for reflection in the classroom and sketch some possible realizations. Prediger (2005b) proposes a framework to characterize possible questions to initiate reflections about mathematics. Chovanetz and Schneider (2008) sketch some elements of possible teaching-learning arrangements that aim at reflection about descriptive statistics; so does Lengnink (2010).

Common to these studies is their character as sketch or proposal. Although they provide theoretically sound reasons, systematic empirical work that realizes their proposals and that provides a detailed account of learners' reflection processes is still underdeveloped. One step towards systematic empirical work is carried out by Kröpfl (2007), who not only provides a normative framework for reflective knowledge regarding the learning content of functions, but also reports on an implementation of a curriculum aiming to realize the learning of reflective knowledge. Kröpfl shows how reflection seems to have taken place during his teachings, but due to methodological restrictions cannot show the conditions and typical pathways of learners' actual formation of reflective knowledge.

The state of theory calls for an existence proof of meaningful reflection in classrooms at a micro level along with a theory-driven analysis of empirical processes detailing how such reflections can be elicited. Thus, apart from the specification gap, a *realization gap* exists in research on reflection in mathematics education concerning the lack of theory and design to initiate and support the development of reflective concepts: a formulation of a hypothetical learning trajectory towards the reflective goals of mathematical literacy in statistics remains impossible without the specification of general reflective concepts to be learned as well as possible situative reflective concepts providing the foundation for learning.

After the previous chapters have introduced a normative framework (Chapter 2) and the specification of the learning content of statistical measures (Chapter 3), this chapter has introduced the learning-theoretical background of this thesis. This background allowed to sketch a general learning trajectory towards the mathematizing goals, but has also identified a realization gap for the reflective goals of mathematical literacy in statistics. These two results conclude the theoretical work of this thesis; the next chapter provides a summary of the work so far and introduces the empirical research questions for the empirical parts of this thesis.

5 Summary and research questions

So far, this thesis has argued that:

Chapter 1. There is a need for the design of a compact teaching-learning arrangement for statistics. This requires careful re-evaluation of the general aims and the specific goals of statistics instruction.

Chapter 2. In order to attain mathematical literacy, learners need to be enabled to reach both, mathematizing and reflective aims of mathematical literacy. For this, they need to reach the mathematizing and reflective goals. For each side of mathematical literacy, these goals consist of developing specific concepts and mastering specific activities. The specification gap concerning reflective concepts however hinders systematic research into students' development of the reflective side of mathematical literacy, and some few reflective concepts as goals can only be specified preliminarily. The mathematizing concept of measure promises to be a useful starting point for the development of both sides of mathematical literacy, because it provides the elementary statistical tools which also play a central role for the relationship between mathematics and society.

Chapter 3. Measures are data-based formalized representations of aspects of phenomena. There exist general measures as well as situative measures. Learners' informal reasoning can be understood as the use of intuitive situative measures. The mathematizing concept of measure needs to be developed before more advanced concepts like informal statistical inference can be developed. However, research so far has not explicitly focused on learners' development of measures and has overemphasized the features of the data identified by learners over aspects of phenomena.

Chapter 4. Mathematizing and reflective concepts are developed while engaging in mathematizing and reflective activity within mathematizing and reflective situations. In learning processes, concepts are situative. The general mathematizing concepts of the mathematizing goals can be developed from learners' intuitive situative measures. For the reflective side however, the realization gap of reflection hinders systematic research into students' development of the reflective side of mathematical literacy, because no hypothetical learning trajectory can be constructed that could inform the design of a teaching-learning arrangement.

© Springer Fachmedien Wiesbaden GmbH, part of Springer Nature 2018
C. Büscher, *Mathematical Literacy on Statistical Measures*,
Dortmunder Beiträge zur Entwicklung und Erforschung des
Mathematikunterrichts 37, https://doi.org/10.1007/978-3-658-23069-2_5

Each chapter however also raised some questions:

Chapter 1. What are the most important learning contents of statistics? How can a suitable teaching-learning arrangement look like?

Chapter 2. How can the mathematizing concept of measure be conceptualized? What are the general reflective concepts making up the reflective goals of mathematical literacy?

Chapter 3. What is a conceptualization of the mathematizing concept of measure? What are learners' intuitive situative measures?

Chapter 4. How do learners develop general mathematizing concepts? What are learners' situative reflective concepts? Is there a connection between mathematizing and reflective activity and therefore between the development of mathematizing and reflective concepts?

Some of these questions have already been answered through theoretical work. The question of important learning contents has been answered through the constructs of the sides of mathematical literacy (Chapter 2). The question of the mathematizing concept of measure has been answered by the conceptualization of measure through the constructs of formal characteristics, features of the data, and aspects of phenomena (Chapter 3). The question of the possible development of general mathematizing concepts through measures has been answered by constructing a hypothetical learning trajectory (Chapter 4).

The remaining questions require empirical insights into students' learning processes. Although a teaching-learning arrangement can be designed in theory, its actual effects need to be judged empirically. Insights into learners' situative concepts and their development require an investigation into their learning processes. And, as argued in the following chapter, empirical work can even support the specification of general reflective concepts. As such, this thesis turns towards the empirical part to answer the following empirical research questions:

RQ1. How can a teaching-learning arrangement support the mathematizing goals of mathematical literacy in statistics?

    a. How can a teaching-learning arrangement initiate and support learners' mathematizing activities?

    b. What is a context for a teaching-learning arrangement that allows learners to develop situative measures through mathematizing activities?

    c. How do learners' intuitive situative measures develop during their mathematizing activities?

RQ2. How can a teaching-learning arrangement support the reflective goals of mathematical literacy in statistics?
  a. How can a teaching-learning arrangement initiate and support learners' reflective activities?
  b. Which situative reflective concepts do learners draw on during their reflective activities?
  c. What are possible general reflective concepts and how do they relate to situative reflective concepts?

RQ3. What is the interplay between learners' mathematizing and reflective activities?
  a. How can learners' mathematizing activities support their reflective activities?
  b. How can learners' reflective activities support their mathematizing activities?

These research questions require empirical work. As such, this PhD-thesis turns towards its empirical part. The following chapter introduces the methodological framework for the underlying Design Research study.

6 Methodology

The empirical research questions formulated in Chapter 5 require a research framework that allows to investigate students' learning processes as they can be initiated by a carefully designed teaching-learning arrangement. For this purpose, the Design Research framework is suitable which is introduced ub Section 6.1. Section 6.2 provides an overview on the participants and the methods of data collection. The method of data analysis is then described in Section 6.3.

6.1 Research framework

The research questions in Chapter 5 call for a framework that allows to combine insights into students' learning processes and the workings of teaching-learning arrangements. Design Research (Gravemeijer & Cobb, 2006) can provide such a framework. This section introduces topic-specific Didactical Design Research as its central methodology (Section 6.1.1). Afterwards, Educational Reconstruction is introduced as an approach to closing the specification gap of reflective concepts (Section 6.1.2).

6.1.1 Topic-specific Didactical Design Research

To answer the research questions, the underlying study of this thesis chose the framework of Design Research which provides an interventionist approach to research (Cobb, Confrey, diSessa, Lehrer, & Schauble, 2003; Gravemeijer & Cobb, 2006; for a larger overview see van den Akker, Gravemeijer, McKennedy, & Nieveen, 2006; Plomp & Nieveen, 2013). Instead of only observing current teaching practice (e.g. Krummheuer & Jungwirth 2006) or 'uncovering' students' reasoning with as little interference as possible by researchers (e.g. Kahneman & Tversky, 1979), Design Research aims at two interchanging goals: (A) the development of empirically validated teaching-learning arrangements via identification of suitable design elements and more general design principles, and (B) contributions to local theories about the initiated teaching learning processes with their typical learning pathways and obstacles as well as insights into the functioning and conditions of success for specific design elements (Prediger, Gravemeijer, & Confrey, 2015).

More precisely, the Design Research project presented here follows the approach of topic-specific Didactical Design Research with a focus on learning processes (Prediger et al., 2012; Prediger & Zwetzschler, 2013). Research in this framework is conducted throughout four working areas (Fig. 6.1). Specific to this approach to Design Research is the working area of *specifying and structuring learning goals and contents*. Building on theoretical considerations as well as empirical observations, this working area aims at identifying relevant

© Springer Fachmedien Wiesbaden GmbH, part of Springer Nature 2018
C. Büscher, *Mathematical Literacy on Statistical Measures*,
Dortmunder Beiträge zur Entwicklung und Erforschung des
Mathematikunterrichts 37, https://doi.org/10.1007/978-3-658-23069-2_6

knowledge about the learning content to be gained by learners (specifying) and finding connections between knowledge elements and sequences of learning (structuring) – a work that has been done in the preceding chapters and continues through the empirical part. Formal, normative, and epistemological considerations all can be part of this working area (Hußmann & Prediger, 2016).

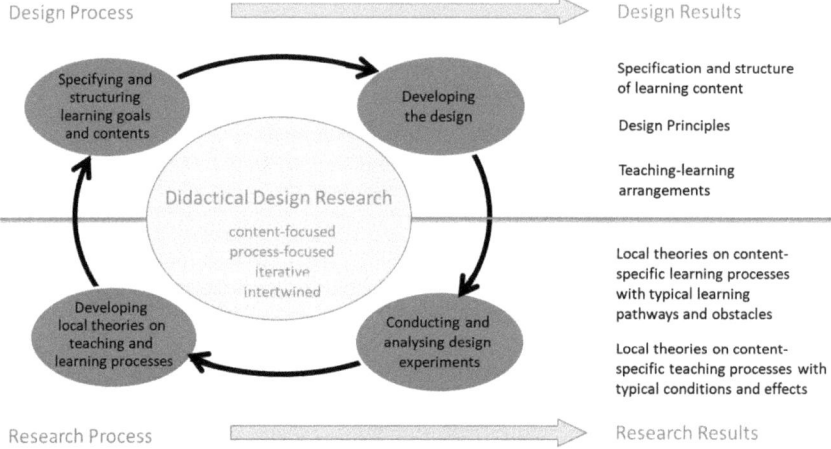

Fig. 6.1: The four working areas of topic-specific Design Research (Prediger et al., 2012; translated in Prediger & Zwetzschler, 2013, p. 411)

The specified and structured learning content informs the working area of *developing the design*. The aim of this working area is the development of a teaching-learning arrangement that elicits processes of learning the structured content. A major part is the identification of *design principles*, i.e. general principles postulating how to reach intended aims or process qualities in the teaching-learning-arrangement (van den Akker, 1999; Gravemeijer & Cobb, 2006). These design principles are implemented by means of concrete *design elements* making up the actual teaching-learning arrangement.

The functioning of the design is empirically examined by *conducting and analyzing design experiments*. Design experiments provide the methodological core of Design Research (Brown, 1992; Cobb et. al., 2003; Gravemeijer & Cobb, 2006). The setting can vary from classroom settings (e.g. Bakker, 2004) to laboratory settings with few students (e.g. Glade, 2016). The scope can vary from single design experiments to series of multiple consecutive design experiments (Cobb et al., 2003). Regardless of setting and scope, the aim is not to simply observe but to actively initiate and interfere with students' learning processes in order to understand their preconditions, typical pathways, and obstacles.

The analysis of the design experiments feeds into the working area of *developing local theories on teaching and learning processes*. The aim is to construct theories that allow to describe and explain students' diverse individual learning pathways through the structured learning content and the effects of the design principles. Those theories, however, stay local in that they do not make universal claims about the nature of students' learning, but are always framed in the design of the teaching-learning arrangement and tied to the specific topic in view (Cobb et al., 2003; Prediger et al., 2015).

These four working areas are not configured in a unidirectional sequence, but rather in iterative cycles of design and research where the working areas are highly intertwined. This allows the study to react to conditions of the field that cannot be predicted through theory, a feature common to most approaches to Design Research (Plomp & Nieeven, 2013). It allows an iterative refinement of design and local theories, resulting in focused theories on learning and design.

Thus, the framework of topic-specific Design Research is suitable for a study addressing the need of development of a teaching-learning arrangement for statistical measures. The learning content of statistical measures on the mathematizing side (Chapter 2) has been specified through the distinction of formal characteristics, features of the data, and aspects of phenomena (Chapter 3). Some structure is given through the development of measures through mathematizing activity (Chapter 4); however, more insights into this development are needed, illustrating how the task of specifying and structuring the learning content also has an empirical part. Likewise, trying to specify and structure the reflective side of mathematical literacy has revealed the specification and realization gaps (Chapters 2 and 4), also requiring empirical work.

6.1.2 Educational Reconstruction of reflective concepts

Due to the existing specification gap of reflection, general reflective concepts cannot be specified based on theoretical research alone. Thus, the identification of relevant reflective concepts has to be treated as an empirical task. This approach has its roots in the framework of Educational Reconstruction (Duit et al., 2012; Kattmann & Gropengießer, 1996; Kattmann, Duit, Gropengießer, & Komorek, 1997).

The main assumption of Educational Reconstruction is that learning contents are not determined by the discipline itself. A 'topic-specific clarification' of learning contents from the perspective of the discipline can only identify the part of the learning content that concerns relevant topic-specific knowledge. Target conceptions to be held by learners, possible conceptions that actually are held by learners and which could support or hinder learning, and insights into possible learning processes, however, also are important part of the learning content. Yet such insights cannot be gained from such a topic-specific clarification alone. The topic-specific clarification has to be complemented by the step of 'collecting and comprehending students' perspectives'. Whereas the former is of theoretical nature, the latter is an empirical task. Only by reconciling theoretical and empirical insights can the actual learning content be specified, i.e. didactically reconstructed.

In Germany, the framework of Educational Reconstruction has been used for the specification of learning contents of a wide range of disciplines and topics (e.g. for stochastics Prediger, 2008; for biology Riemeier & Gropengießer, 2008). In this study, concerning the mathematizing side of mathematical literacy, the learning content of statistical measures has already been didactically reconstructed by reconciling a topic-specific clarification (Chapter 2) with the collection and comprehension of students' perspectives (Chapters 3 and 4) – although Educational Reconstruction can never be said to have been completed, and more empirical insights are needed. For the reflective side of mathematical literacy, theoretical considerations have revealed some few possible general reflective concepts. The major empirical task of specifying further general reflective concepts and of identifying their situative precursors is still outstanding.

6.2 Methods of data collection

In line with the qualitative nature of design research, data has been collected from students' learning processes during the design experiments. The study began in March 2014 with theoretical preparations. Four cycles of design experiments have been conducted from June 2014 to December 2015.

Cycles and participants

Table 6.1 presents an overview of the cycles of design experiments. Each cycle consisted of several design experiment series ranging from one (in Cycle I) to three (in Cycle IV) consecutive design experiment sessions in a laboratory setting. During the design experiments, pairs of students worked on one teaching-learning arrangement under supervision of the design experiment leader (the author). The reason for this setup was that in pairs, students could interact with each other instead of only with the design experiment leader. In this way, they could share ideas with each other and engage in mathematizing or reflective activity together, creating richer learning processes that revealed more of the students' reasoning. To further support these rich activities, the design experiment leader took care that the students knew the purpose of the design experiments. It was emphasized that the students were not placed under test conditions, but that the aim was to evaluate the teaching-learning arrangement and to understand the students' thinking. The students were encouraged to articulate their thoughts even when they were unsure.

Tab. 6.1: The cycles of design experiments of this study

Cycle	Time	# students	# design experiments	Total video
I	Jun. 2014	6	3	~ 135 min.
II	Nov. 2014	10	6	~ 340 min.
III	May 2015	10	10	~ 450 min.
IIIb	Oct. 2015	4	4	~ 180 min.
IV	Dec. 2015	6	9	~ 405 min.
Total		34	32	~ 1510 min.

Each cycle took place in a different school in Germany (with the exception of Cycle IIIb, which took place in the same school as Cycle II). The students participating in the design experiments were sampled from 7th Grade mathematics classrooms of those schools. This grade was chosen for reasons of syllabus fit, as in Germany, statistics education takes place within mathematics classroom, and measures of spread are a learning content for 7th or 8th Grade. For sampling, the teachers of 7th Grade were asked to choose students exhibiting good to average mathematics abilities and high communication skills. Thus, the sample was not random and consisted of students volunteering to take part in the experiments. Due to the explorative nature of this study, this allowed a more focused approach and richer discussion to be analyzed. Attention was given to that the students had not had statistics lessons for 7th Grade yet, in order to assure the comparability of students' former experiences between cycles. Individual statistical pre-knowledge was not controlled. The syllabus suggests that the students knew the mean as a method of summary, but not as a measure of center and that the students were familiar with bar and pie charts, but not with histograms or dot plots.

As the design evolved over the cycles, so did the focus of the study. This thesis reports on results of the later design experiment cycles, as the development of measures only became the focus of the study with Cycle III. Design and local theories from Cycles I and II are not explicated in this PhD thesis, but only serve as underpinnings of this thesis (for results of Cycle I and II see Büscher, 2015; Schnell & Büscher, 2015).

Methodological control of roles of the design experiment leader

The presence of the design experiment leader during the design experiments allowed to intervene with the students' learning processes by prompting them to verbalize their thoughts and or to pose challenging questions. When giving these prompts, the design experiment leader fulfilled two different roles:

• in the role of the teacher, the design experiment leader aimed at eliciting activities that supported the students' development of measures;

- in the role of the researcher, the design experiment leader aimed at uncovering the students' own reasoning with as little interference as possible. In each design research project, the interplay between both roles must be carefully controlled in order to not to get into conflict. This was realized by a semi-standardized protocol which included prepared prompts for specific anticipated situations during the design experiments. The protocol is presented in more detail for Cycle III in Section 7.1.

Video and screen capture data

The laboratory setting provided a semi-standardized setting in which students, task, and design experiment leader were the main factors. Because of the richness of learning processes, the design experiments were completely videotaped to allow a fine-grained analysis of students' verbal expressions and gestures on a micro level that would not have been possible by only taking field notes. As Table 6.1 shows, in total about 1510 minutes (= 25 hours) of video data were gathered.

Additionally, all students' written products and screen capture data belong to the data corpus: all statistical data used in the design experiments were presented to the students via TinkerPlots 2.0 (Konold & Miller, 2011) running on a tablet computer (TinkerPlots' features, however, were made unavailable to the students, and the students did not actually make use of the software for the problem). In order to capture students' interaction with the data, a screen overlay software running on the tablet enabled the students to use the tablet's digital pen to draw freely anywhere on the tablet screen. This allowed them to mark features of the data relevant to them and to illustrate situative measures which did not conform to the pre-made tools in TinkerPlots. A screen capture software recorded these markings, which could then be used in data analysis.

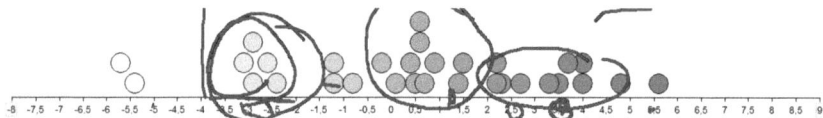

Fig. 6.2: Students' structuring of the data through the overlay software

Figure 6.2 gives an example in which two students used the digital pen to mark different intuitive features of the data. The students seem to identify a cut-off point which separates the 'main' body of data from the 'exceptions' (the vertical line at -4). They also identify three clumps which seem to hold different meanings to the students (the three circles around the data). Additionally, some specific values seem to be of special interest, possibly for representing the various clumps (the numbers -3, 2.5, 3 are marked). These marked features of the data, however, are intuitive and do not correspond neatly to any general mathematizing concept. Only the screen overlay and capture software enabled the students to express their thinking. Thus, this method provides valuable insights into how the students engage in structuring data.

6.3 Methods of data analysis

A non-reductionist analysis of learning processes that does justice to the richness of students' reasoning expressed through various situative concepts-in-action cannot be accomplished for all of 1510 minutes of video data for the whole project, or even for the 855 minutes of Cycles III and IV, thus the data had to be reduced to manageable size, giving careful consideration to retaining the richness of the learning processes.

The further methodological considerations have been shaped by the chosen learning-theoretical background provided by Vergnaud (1996) and Pratt and Noss (2002). They propose that students' individual intuitive reasoning does not necessarily conform to formal ideas and is heavily influenced by the actual situations of their learning processes. In order to give justice to the situative nature of students' reasoning, the data has to be analyzed not by mapping students' whole learning processes to few normative general concepts, but rather in a way that retains the individual, subjectively meaningful nature of students' situative concepts-in-action. In order to take these methodological aspects into consideration, the analysis of Cycles III and IV consisted of several steps.

Step 1. Selecting focus pairs and creating transcripts

As a first step, the whole corpus of video data was reviewed, and rough summaries of the design experiments were created in order to provide an overview of the learning processes. These summaries showed differences in richness of activity for different pairs of students. Four pairs stood out with respect to the intensity of their interaction, explicitness in verbal reasoning, and richness of activity: For Cycle III, Maria & Natalie and Quanna & Rebecca (see Chapter 7); for Cycle IV, Kaan & Nesrin and Jana & Mara (see Chapters 8 and 9). The students were chosen not because of any exceptional achievements during the design experiments, but because of their communicative strength allowing greater insights into their learning processes.

For these selected focus pairs chosen for the qualitative in-depth analysis, complete transcripts of their design experiments were created (altogether transcripts for ~ 450 minutes of video). Transcripts are crucial for the qualitative analysis as they provide a linear text-based structure. Students' German verbal verbal expressions were transcribed in German as faithful as possible to their actual use of language, with some orthographic additions to increase readability (i.e. adding full stops and commas when needed). (The translation into English was only conducted after all other steps of analysis in Step 8). Students' gestures were also transcribed when necessary to understand their verbal verbal expressions. This proved to be extremely important, as having only developed a yet limited statistical language, the students made frequent use of deictic language means when referring to data. Apart from the students' verbal expressions, the transcripts made use of some special characters: square brackets [] for gestures and nonverbal communication, parenthesis () for measured pauses, ellipses … for unmeasured pauses and communication trailing off, and '#' for concurrent speaking.

For the pairs of students not identified as focus pairs, some excerpts of the design experiments were nevertheless useful for analysis and were transcribed accordingly. These non-focus pairs provided comparative cases for the later steps of analysis. Although their learning processes were not analyzed in detail, the emerging categories of the analysis (see Step 4) were tested against these non-focus pairs. This ensured that the categories were gainfully applicable to more than only the special cases of the focus pairs, and thus created a higher level of generalizability for the studies' results.

The transcripts of the focus pairs' learning processes then could be investigated on a micro-level. Still, the data had to be reduced further.

Step 2. Identifying sequences relevant to research questions

Within the transcribed 450 minutes of design experiments for the focus pairs, all sequences relevant to the research questions were identified in Step 2 of the analysis. As Cycle III aimed at investigating the students' use of 'typical' (see Chapter 7), sequences were focused in which the students explicitly or implicitly referred to measures or ideas concerning 'typical'. Cycle IV investigated students' reflective activities, so that sequences were focused in which students did not only use or develop measures in mathematizing activity, but in which they referred to the use or development of measures on a meta level – i.e. in which they were in situations of advocating or rejecting measures (see Chapter 8).

These sequences then provided the objects of the actual micro-level analysis.

Step 3. Basic reconstructive analysis of concepts-in-action and theorems-in-action

With the relevant sequences identified, the qualitative in-depth analysis of the students' learning processes on a micro level could proceed with the basic analysis which provided the foundation for all following steps of analysis. For capturing the situative and highly individual nature of students' first ideas, the analytic procedure had to unpack students' individual perspectives rather than just coding with a pre-existing coding scheme. Within the German methodological tradition, these interpretations of individual sense-making are called *reconstruc-*

tive (e.g. Bohnsack, 2008); the verb *reconstructing* is used to signify this constructive act of comprehending students' subjective and internal logic.

For operationalizing the basic analysis, an analytic approach based on Vergnaud's (1996) constructs of theorems-in-action and conceptions-in-action as applied in the adaptations by Glade & Prediger (2017; similar Prediger & Zindel, 2017; Zwetzschler, 2015; Glade, 2016). Vergnaud describes his epistemological theory of learning as "a fruitful and comprehensive framework for studying complex cognitive competences and activities and their development" (Vergnaud, 1996, p. 219). He assumes that for each situation, individuals develop specific or more general schemes of action to cope with these situations. These schemes are, implicitly or explicitly, guided by so-called situational invariants, the theorems-in-action and concepts-in-action. Vergnaud defines a *theorem-in-action* as "proposition that is held to be true by the individual subject for a certain range of situation variables" (Vergnaud, 1996, p. 225), so the analytic approach reconstructs the implicit individual propositions that guide the students' verbal expressions and actions. These theorems-in-action are shaped by *concepts-in-action,* defined as "categories (...) that enable the subject to cut the world into distinct (…) aspects and pick up the most adequate selection of information" (ibid.). To stay as close to the students' reasoning as possible, their own use of language was used for naming the concepts-in-action and theorems-in-action where possible in this first step.

The following transcript serves as an example of a (very short) sequence which was chosen for the student Nesrin's use of meta language about the use of measures (excerpt from Chapter 8). The transcript is preceded by a short description, detailing the cycle of design experiments (IV), the session of the design experiment series (2), the initials of both students (KN), the time in the design experiment (20:10) and short description of what happened previously and what happens in this scene. In this transcript, Nesrin refers to different situative measures (Most Important Value and Distance) of the phenomenon of Arctic sea ice (see Chapters 7 and 8).

IV-2-KN; Phase 3; Start: 20:10
Previously, Nesrin identifies the new aspect of the point where the ice goes back to. Now, she resolves the conflict between Distance and Value Report Sheets by differentiating between different aspects of the phenomenon to be addressed by the different measures.

| 137 | N | So, um, if you – here *[Value Report Sheet]* do you have the 16 and there the 15 *[Value Report Sheet]* and when you look at this one *[hints at Distance Report sheet],* they say that it has gone back by half. And here they say *[hints at Value Report Sheet]* nothing about decreasing and, or about going up, they only say that in the end the ice was that big again. | Also ähm, wenn man halt – hier *[deutet Wert-Steckbrief]* hat man ja die 16 und da die 15 *[deutet Wert-Steckbrief]* und wenn man das aber in dem hier guckt *[deutet auf Abstand-Steckbrief],* da sagen die, dass es um die Hälfte zurückgegangen ist. Und hier gar *[deutet auf Wert.-Steckbrief]* sagen die ja nichts mit zurückgehen und oder wieder nach vorne gehen, |

			die sagen nur am Ende war das Eis wieder so groß
138	DL	Mhm.	Mhm.
139	N	And I think they could also rather have said what happened in the middle.	Und die hätten glaub ich auch ruhig sagen können, was in der Mitte passiert ist.

Several theorems-in-action (marked by $<...>$) were reconstructed from this transcript. These were closely connected to the students' use of language, resulting in theorems-in-action such as *<the 15 and 16 are important>*, *<the Distance Report Sheet says that the ice has gone back by half>*, and *<in the end, the ice was big again>*. These theorems-in-action draw on concepts-in-action (marked by $||...||$) such as $||the\ ice||$, $||the\ Distance\ Report\ Sheet||$, and $||big\ again||$.

In this step, concepts-in-action regarding the mathematizing side could easily be reconstructed. For the reflective side however, the specification and realization gaps hindered an easy reconstruction of possibly relevant concepts-in-action. To make progress, the analysis used 'sensitizing concepts' to provide a search direction for possible reflective concepts-in-action. These were the reflective concepts already identified in Chapter 2: manipulation, contextual relevance, and perspectivity. However, The coding was still done in an open way to permit other types of reflective concepts-in-action to be reconstructed.

These first reconstructed concepts-in-action and theorems-in-action only marked the beginning of the analysis, supplying the foundation for the step of inductive category development.

Step 4. Inductive category development

For structuring the reconstructed concepts-in-action and theorems-in-action named by drawing on students' language in the basic analysis, a process of inductive category development similar to the corresponding stages in grounded theory (Corbin & Strauss, 1990) or 'inductive category formation' in qualitative content analysis (Mayring, 2015). By comparing and contrasting the reconstructed concepts-in-action and theorems-in-action, similarities and differences between the students' reasoning could be identified and successively better grasped by the categories. The emerging categories were refined through iterative re-formulating of the students' concepts-in-action and theorems-in-action and further comparing and contrasting.

For distinguishing the theorems-in-action and concepts-in-action, it turned out to be relevant to distinguish

- the mathematizing or reflecting activity, in which the situative theorems-in-action were reconstructed ('structuring', 'representing', 'formalizing', 'identifying patterns of thought', 'explicating aims and purposes', 'denominating risks and limits'; see Section 2.2.2 for mathematizing activities and Section 2.3.2 for reflective activities)

- the different types of elements to which a concept-in-action refers ('reflective concept' if some meta-aspect was addressed or the following reference for mathematizing concepts: 'formal characteristic', 'feature of the data', 'aspect of the phenomenon'; see Chapter 3)

These constructs of 'reflective concepts', 'reflective and mathematizing activities', 'features of the data', and 'aspects of phenomena' can be considered *ontological innovations* (diSessa & Cobb, 2004). They were created through the category-developing approach because they demonstrated high explanatory power for the complex learning processes observed in the design experiments. Ontological innovation is one way how Design Research fulfills a theory-generating role (Prediger et al., 2015). These constructs could not have been foreseen without the empirical work, and present major theoretical results of this study; still, in order to instantiate them as genuine theoretical constructs, they needed to be theoretically embedded in literature (diSessa & Cobb, 2004). Thus, Chapters 2 to 4 of this PhD thesis already are the result of the empirical Design Research that was carried out.

The developed category-system could then be utilized to categorize the reconstructed concepts-in-action and theorems-in-action. The rules for assigning categories build on the definitions presented in Section 4.1.3. Table 6.2 gives an overview on all categories for the theorems-in-action and concepts- in-action.

Tab. 6.2: Rules for assigning categories to reconstructed concepts-in-action and theorems-in-action in-action

Abbreviation	Category	Explanation
MA -	**Categories for theorems-in-action according to the mathematizing activities in which they emerge**	
MA/s	Structuring	Theorem-in-action reconstructed in situations of predicting phenomena or ensuring communication, concerning only concepts-in-action of aspects of the phenomenon, features of the data, or the relation between these categories.
MA/r	Representing	Theorem-in-action reconstructed in situations of predicting phenomena or ensuring communication, relating concepts-in-action of measures and aspects of the phenomenon or features of the data.

Tab. 6.2: Rules for assigning categories to reconstructed concepts-in-action and theorems-in-action in-action (continued)

Abbreviation	Category	Explanation
MA/f	Formalizing	Theorem-in-action reconstructed in situations of predicting phenomena or ensuring communication, relating concepts-in-action of measures and formal characteristics.
RA -	**Categories for theorems-in-action according to the reflecting activities in which they emerge**	
RA/i	Identifying patterns of thought	Theorem-in-action reconstructed in situations of advocating or rejecting measures, relating concepts-in-action of measures and reflective concepts-in-action.
RA/e	Explicating aims and purposes	Theorem-in-action reconstructed in situations of advocating or rejecting measures, relating concepts-in-action of measures and their utility through a reflective concepts-in-action in order to increase or justify utility.
RA/d	Denominating risks and limits	Theorem-in-action reconstructed in situations of advocating or rejecting measures, relating concepts-in-action of measures and their utility through a reflective concepts-in-action in order to limit utility.
MC -	**Categories for the references of the mathematizing concepts-in-actions**	
MC/m	Measure	Concept-in-action concerning formalized representations of aspects of phenomena; used in theorems-in-action concerning representing and formalizing.
MC/c	Formal characteristic	Concept-in-action concerning mathematical properties that are not directly represented in data or phenomenon, such as form and procedure of calculation; used in theorems-in-action concerning formalizing.
MC/f	Feature of the data	Concept-in-action concerning the data without contextual interpretation, such as location or frequency; used in theorems-in-action concerning structuring and formalizing.
MC/a	Aspect of the phenomenon	Concept-in-action concerning the partition of phenomena using contextual knowledge; used in theorems-in-action concerning structuring and representing.
	Further categories for concepts-in-action	
RC	Reflective concept	Non-mathematizing concept-in-action in a theorem-in-action concerning reflective activity, addressing a meta-aspect; reconstructed in situations of advocating or rejecting measures.
MCs / RCs	Situative concept	Concept-in-action situated in specific contextual neighborhoods and bound to concrete phenomena.
MCg / RCg	General concept	Concept-in-action not situated in specific contextual neighborhoods and bound to general learning situations.

In the following step, the rules for assigning categories were used to analyze the students' mathematizing pathways.

Step 5. Sequential analysis of mathematizing pathways

Step 3 and 4 allowed now the most crucial step, the sequential analysis of students' mathematizing pathway. The same transcript as above now serves as an example of the deductive use of the developed categories.

IV-2-KN; Phase 3; Start: 20:10
Previously, Nesrin identifies the new aspect of the point where the ice goes back to. Now, she resolves the conflict between Distance and Value Report Sheets by differentiating between different aspects of the phenomenon to be addressed by the different measures.

137	N	So, um, if you – here *[Value Report Sheet]* do you have the 16 and there the 15 *[Value Report Sheet]* and when you look at this one *[hints at Distance Report sheet]*, they say that it has gone back by half. And here they say *[hints at Value Report Sheet]* nothing about decreasing and, or about going up, they only say that in the end the ice was that big again.	Also ähm, wenn man halt – hier *[deutet Wert-Steckbrief]* hat man ja die 16 und da die 15 *[deutet Wert-Steckbrief]* und wenn man das aber in dem hier guckt *[deutet auf Abstand-Steckbrief]*, da sagen die, dass es um die Hälfte zurückgegangen ist. Und hier gar *[deutet auf Wert.-Steckbrief]* sagen die ja nichts mit zurückgehen und oder wieder nach vorne gehen, die sagen nur am Ende war das Eis wieder so groß
138	DL	Mhm.	Mhm.
139	N	And I think they could also rather have said what happened in the middle.	Und die hätten glaub ich auch ruhig sagen können, was in der Mitte passiert ist.

The analysis of the students' mathematizing pathways started by identifying students' mathematizing activities based on their theorems-in-action (indicated by $<...>$, preceded by 'MA/s' for structuring, 'MA/r' for representing, and 'MA/f' for formalizing) and their mathematizing concepts based on their concepts-in-action (indicated by $||...||$, preceded by 'MCs' for situative and 'MCg' for general concepts). The mathematizing concepts are then assigned to the categories of measures, formal characteristics, features of data, or aspects of phenomena.

In the above example, Nesrin refers to "the ice in the end" (#137) and "what happened in the middle" (#139). In the analysis, this is interpreted as referring to a 'recovery point' and the 'melting process' of Arctic sea ice (see Chapter 8) – which only became visible through the analysis of the whole pathway through the design experiment, instead of the few isolated lines of the transcript. Because this refers to the phenomenon, her verbal expressions are coded as the concepts-in-action of aspects of the phenomenon $||MCs/a: recovery point||$, and $||MCs/a: melting process||$. These concepts-in-action are utilized by Nesrin in a theorem-in-action that shows her structuring activity: $<MAs: Arctic sea ice is composed of the ice in the end and what happened in the middle>$. She also refers to a situative measure of $||MCs/m: Distance||$, which she uses to address the aspect of the $||MCs/a: melting process||$ (by referring to the 'Distance Report

Sheet', see Chapter 7). This is coded as a theorem-in-action concerning the mathematizing activity of representing ('MA/r'): $<MA/r$: *The Distance represents the melting process>*.

Step 6. Analysis of reflective activities and theorems-in-action

The reflective theorems-in-action are reconstructed in the basic analysis in Step 3 and categorized according to the reflecting activities in Step 4. For example, in the exemplary transcript in view above, Nesrin formulates a clear preference for the use of the Distance over the Most Important Value (#139). Nesrin judges the usefulness and adequacy of the Distance for Arctic sea ice. This allows to reconstruct her theorem-in-action in the reflective activity of explicating aims and purposes ('RA/e'): $<RAe$: *the Distance should be used instead of the Most Important Value for arguing about Arctic sea ice>*.

Step 7. Analysis of reflective concepts

In the last step, the situative ('RCs') and general ('RCg') reflective concepts-in-action which are latent in the students' reflective theorems-in-action were identified. Due to the specification and realization gaps for reflective concepts in the current state of research and the limited occurrences of reflective concepts-in-action, no further categorization was conducted. A further structuring of the situative reflective concepts with respect to potential general reflective concepts is presented as the final outcome of the comparison and contrasting procedure in the empirical Chapter 8. Only by specifying for which general reflective concept a given situative reflective concept is a precursor to can both concepts reciprocally gain meaning as *reflective* concepts.

A coding of situative and general reflective concepts thus has two functions: (1) the proposal for that general concept to be considered a canonical reflective concept in an attempt to specify general reflective concepts needed to attain the reflective goals, and (2) a bold interpretation of the situative reflective concept-in-action as being a precursor to this general reflective concept in an attempt to uncover pathways for development of situative to general reflective concepts.

In the above example, Nesrin gives preference a specific aspect. She compares the two measures and their represented aspects after $||RCs$: *what one should say*$||$, possibly a situative starting point for $||RCg$: *contextual relevance*$||$: not only do measures only address specific aspects, but one has to decide which aspect actually should be addressed.

The results of this analysis are then summarized in diagrams, showing the whole conceptual network of the students until this point of the design experiment. This allows to quickly overlook the state and changes of students' learning processes (Fig. 6.3)

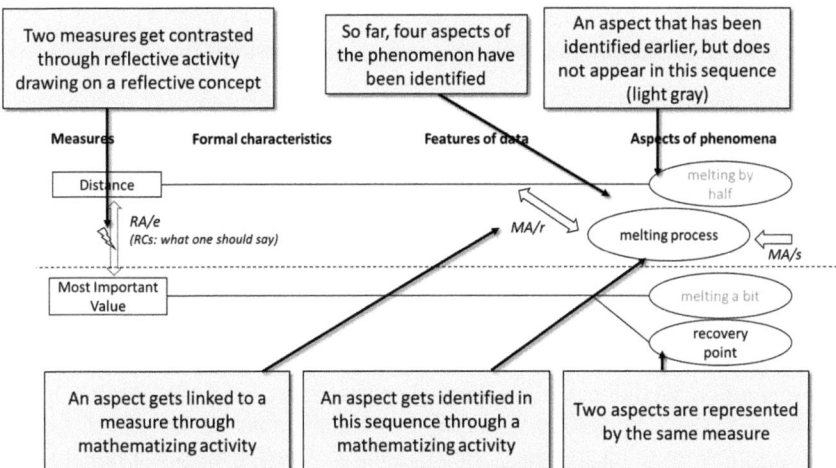

Fig. 6.3: Diagrams in the analysis provide an overview about the students' whole conceptual network as well as current activities

Figure 6.3 does not have to be read in any specific direction. It shows that Kaan and Nesrin so far have addressed two measures (Distance and Most Important Value, separated by a dotted line) and four aspects of phenomena (melting by half, melting process, melting a bit, and recovery point). Two of these aspects of phenomena are relevant to the current sequence (melting process and recovery point), the others are not explicitly referred to. The aspect 'melting by half' was represented by the students through use of the measure Distance in a previous sequence (indicated by a single black line). The aspect 'melting process', however, is newly identified through the mathematizing activity of structuring (indicated by double arrows with 'MA/s'). It is also immediately used in representing activity concerning Distance (double arrows with 'MA/r' pointing on the previously established connection between Distance and 'melting by half'). The Most Important Value was linked to two aspects of the phenomenon in a previous sequence (indicated by single black line). Additionally, the students contrast the two measures (indicated by dotted double arrows with lightning symbol) by engaging in the reflective activity of explicating aims and purposes ('RA/e'), drawing on the reflective concept of 'what one should say' ('RCs'). Such a diagram is created for every sequence of the micro-analysis.

Step 8. Translation into English

As design experiments took place in German schools and German is the first language of the author, analysis was conducted in German using German transcripts. This thesis, however, aims at an international audience, so that transcripts and results of the analysis had to be translated to English. This, however, posed critical challenges of translation, as the following transcript illustrates.

III-1a-MN; Starting 14:55
Natalie is comparing the temperature data between two years.
217   N   Yeah, okay. I'd say they are almost   Ja, ok. Ich würde mal sagen, die sind doch
              equally spread out.   fast gleich verteilt.

In German, Natalie uses the expression 'gleich verteilt sein' to compare two data sets. Literally translated, this expression would mean "to be distributed equally". This translation, however, would also transform the expression into a more formal register of language. The expression 'gleich verteilt sein' shows the same root as 'Verteilung', the German word for distribution; as a single word, 'gleichverteilt' would mean the formal characteristic of a distribution being uniform. It is safe to assume, however, that Natalie does not have such formal considerations in mind. Instead, 'verteilt sein' in German carries a much more informal connotation than 'to be distributed' does in English. A translation of meaning results in the more apt translation of 'spread out' – but it hides the anchoring of the concept of distribution in colloquial German and the implications this could have on learning processes.

With these and additional difficulties like grammatical incongruences between languages in mind, the translation of the analysis was created in cooperation of the author and bilingual speakers. Careful attention was given to translating the informal, colloquial, and oftentimes syntactically incorrect nature of students' spoken language. When in doubt however, translations were preferred that retained the results of the analysis conducted in German. In some cases, this could have resulted in English transcripts showing unnatural student language. This, however, was necessary in order to give a faithful report of the results of the analysis.

This analysis provided detailed insights into the students' learning processes and the impact of the teaching-learning arrangement. The next two chapters present the design of the teaching-learning arrangements for Cycles III and IV as well as the results of the analysis.

7 Developing Typical

The empirical research questions formulated in Chapter 5 concern the effects of a teaching-learning arrangement on (RQ1) the mathematizing side and (RQ2) the reflective side of mathematical literacy, as well as (RQ3) the interplay of mathematizing and reflective activity. The three empirical Chapters 7, 8, and 9 each provide some elements of answers to the research questions. None of those chapters, however, can be precisely mapped to a single research question. Instead, Chapter 7 mainly focuses on the starting points of learners' development by investigating their use of intuitive situative mathematizing concepts. However, their reflective activities and concepts also play important roles in their learn-ing processes. Chapter 8 mainly focuses on learners' situative and general reflective concepts, which cannot be understood without taking their mathematizing activities into account. The relationship between situative and general mathematizing concepts is subject of Chapter 9.

Common to all three empirical chapters and central to the research questions is the importance of the teaching-learning arrangement. Each chapter begins with an introduction of the design principles and teaching-learning arrangements employed in the corresponding design experiments. In this chapter, the design principles for Cycle III are introduced (Section 7.1.1), which the inform the design of a teaching-learning arrangement (Sections 7.1.2 and 7.1.3). The following empirical analysis illustrates students' development of situative measures (Section 7.2), followed by the identification of some overarching phenomena observed during the students' learning processes (Section 7.3).

7.1 Design of a teaching-learning arrangement for Cycle III

Chapter 6 introduced the identification and evaluation of design principles as a central goal of design research. In Section 7.1.1, the central design principles for Cycle III are introduced, which then inform the design of the compact teaching-learning arrangement (Section 7.1.2 and 7.1.3).

7.1.1 Design principles for initiating mathematizing activity

As mentioned in Chapter 6, the identification and consolidation of design principles is a central task in design research. Such design principles provide general heuristics for the design of teaching-learning arrangements. Van den Akker (1999) proposes a general framework for the structure of design principles:

"If you want to design intervention X [for the purpose/function Y in context Z], then you are best advised to give that intervention the characteristics A, B, and C [substantive emphasis], and to do that via procedures K, L, and M [procedural emphasis], because of arguments P, Q, and R" (van den Akker, 1999, p.9).

© Springer Fachmedien Wiesbaden GmbH, part of Springer Nature 2018
C. Büscher, *Mathematical Literacy on Statistical Measures*,
Dortmunder Beiträge zur Entwicklung und Erforschung des
Mathematikunterrichts 37, https://doi.org/10.1007/978-3-658-23069-2_7

Van den Akker proposes a very general structure of design principles mostly aimed at the design of large-scale interventions focusing on curricular development. This structure alone is too general for the topic-specific details in the design of a teaching-learning arrangement. In the current study, design principles had to be formulated in a way that focuses on specific parts of the specified learning content: Eliciting mathematizing or reflective activity, focusing on specific situative concepts, or choosing data that exhibit certain features can all serve as smaller purposes regarding the learning content of statistical measures. Thus, the topic-specific design principles adapt van den Akker's (1999) structure: *For the purpose P concerning the learning content of statistical measures, the design of a teaching-learning arrangement should be given the characteristic C, because of argument A.*

In the following, the three main design principles of Cycle III are illustrated, followed by a short list of additional design principles that are not fully investigated here. These design principles do not provide generally valid statements about guiding learning processes, but rather take the form of hypotheses. They are theoretically motivated and grounded in literature, and have been applied in similar forms in other empirical studies. Nevertheless, they still need to be tested empirically for the topic of statistical measures. This empirical work is carried out in Section 7.2 and Chapter 8.

Drawing on the epistemic function of context (DP/Context)

Many mathematical teaching-learning arrangements utilize some kind of real or imagined situation as context for the problems to be worked on. For statistics in particular, context is held as central to any statistical learning (e.g. Pfannkuch, 2011). A common requirement for a context is the use of real data: the GAISE Report (ASA, 2016) emphasizes the importance of real data and recommends for teachers to "use real data from studies to enliven your class, motivate students, and increase the relevance of the course to the real world" (ASA, 2016, p. 17). Real data here primarily serves as a motivating factor for the students, illustrating how context can serve a *motivating function.*

Context, however, can also have other didactical functions (Winter, 1981). Real-data contexts not only serve as a means to motivate students, but learning about real-world contexts can be an end in itself. Context can have an *empowering function* so that students learn to understand and to act on the world surrounding them (e.g. Skovsmose, 1998). A third function has been thoroughly elaborated by RME Theory (van den Heuvel-Panhuizen, 2001). A context that is relatable to the students can provide the intuitive pre-understandings of mathematical concepts that make up the foundation for concept development. This *epistemic function* of context allows students to build on their own understandings.

The epistemic function of context thus plays an important part in allowing students to identify aspects of phenomena and thus in supporting their mathematizing activities. This leads this thesis to the identification of a first design principle: *To enable concept-development based on learners' situative mathematizing concepts, choose a phenomenon familiar to the students for the context of the problem, because this can enable them to engage in the mathematizing activity of structuring ('DP/Context').*

Enabling choice between different measures (DP/Measures)

Learners need to engage in mathematizing activity in order to reach the mathematizing goals of mathematical literacy (see Chapter 3). This activity, however, needs to show a certain richness: a single aspect of a phenomenon, feature of the data, or formal characteristic identified would not suffice. Instead of holding on to a single measure, learners need to acknowledge that there also exist different ways to structure phenomena, represented by different measures. One way to facilitate this insight is to provide learners with the perspectives and measures of other learners. One example for such an approach is the study by Watson (2007), who introduced cognitive conflict about learners' conception of average by providing them with interpretations that differed from their own. This prompted them to reconsider their own conceptions. In a similar vein, Schnell (2013) describes a teaching-learning arrangement in which pre-formulated student prompts are used to confront the students with other conceptions of stochastic phenomena. For the learning content of statistical measures, a similar effect could be achieved by providing learners with a variety of measures and interpretations so that they, after having engaged in rich mathematizing activity, can consciously choose the measure to be used. This then leads to the second design principle: *To increase the conceptual richness of learners' mathematizing activity, put them into a position that enables them to choose between different measures, because making such a choice can prompt further mathematizing activities of structuring and representing ('DP/Measures').*

Problematizing formalizations (DP/Formalizations)

Whereas the design principles DP/Context and DP/Measures focus on eliciting mathematizing activities of structuring and representing, the activity of formalizing also plays an important role. Learners need to progress from their intuitive, informal ideas towards the use of general measures (see Chapter 4). A central part constituting general measures are their fixed formal characteristics. Identifying formal characteristics, however, is commonly seen as a challenging step for students mostly situated near the end of their learning processes. Formalization thus needs to be explicitly initiated. One such approach is described by Glade (2014), who supports students' processes of 'progressive schematization' through careful prompts and task design. This leads his students to formalize the part-of-part determination of fractions. As with the case of DP/Measures, a single act of formalizing an intuitive situative measure does not suffice. Challenging students' formalizations and providing alternatives could also initiate richer activities of formalizing. This leads to the third design principle: *To support the identification of formal characteristics, explicitly call for and challenge students formalizations, because this can prompt further mathematizing activity of formalizing ('DP/Formalizations').*

Additional design principles

In addition to the three design principles for eliciting mathematizing activity, several other design principles have been identified and realized in the teaching-learning arrangement. Because they are not part of the detailed empirical analysis, they are only briefly introduced here.

- *To enable pointed discussion about measures, provide a materialized form as a scaffold that requires students to explicate their situative measures, because this can prompt them to commit to structure and formalizations which can then become the object of discussion ('DP/Scaffold').* Scaffolding a much-employed design principle in mathematics education that consists in providing supporting elements to learners to enable learning, which later can be removed once learners have acquired the corresponding skill (Wood, Bruner, & Ross, 1976). This design principle proved necessary as Cycles I and II showed how students drew on a rich repertoire of situative concepts, but the lack of incentive to articulate and write them down caused these concepts to quickly fade in and out of discussion (see Büscher, 2015 or Schnell & Büscher, 2015 for design experiments missing this principle).

- *To enable students to articulate their situative concepts, provide them with a repertoire of formal and informal language means, because they are not used to statistical discussion and lack of language means can inhibit discussion ('DP/Language').* The students in Cycle I and II showed a rich repertoire of nonstandard language (similar to Makar & Confrey, 2005). The students, however, were frequently searching for words, and showed signs of frustration when unable to articulate their understandings. Thus, it seemed necessary to support students in their use of language (similar to Prediger & Zindel, 2017). This design principle aims at these language obstacles.

- *To increase the focus on aspects of phenomena and to elicit the mathematizing activity of structuring, subsequently increase the amount of available data, because this can aid the students in distinguishing between random noise and stable aspects of phenomena (DP/Data).* This design principle is similar to the 'growing samples' heuristic introduced by Bakker (2004), in which students are confronted with samples of phenomena with increasingly larger sample size. The design principle DP/Data, however, does not require the use of samples, but allows for any increase in data, so that the teaching-learning arrangement can provide non-random data exhibiting some specific intended features of the data.

- To allow students to use their intuitive situative measures, utilize technology in a way that allows a variety of informal treatments of data, because a rigid tool consisting of only formal procedures and general measures can suppress the students' creative use of situative measures ('DP/Technology'). As the goal of the teaching-learning arrangement is not for the students to achieve competence with professional tools, but to develop mathematizing and reflective concepts, the technology employed needs to support concept development. The use of technology can significantly support students' learning processes in statistics (Biehler et al., 2013). However, students' situative concepts do not necessarily correspond to the general concepts implemented through software, and the choice of a specific representational medium can directly affect students' use of diagrams and their learning processes (Harradine & Konold, 2006) This means that a tool should be used that does not simply provide the students with a fixed array of general measures to be used. Instead, it needs to provide the students with means to express their situative measures even when they do not show any similarities to general measures.

The teaching-learning arrangement that was used in Cycle III implements the design principles in two different problems. Students first engage in the Typical Antarctic Temperatures Problem, followed by the Arctic Sea Ice Problem. Each problem is designed for a duration of about 45 minutes (a typical duration for a lesson in German schools).

7.1.2 The Typical Antarctic Temperatures Problem

The Typical Antarctic Temperatures Problem introduces the central design elements (stacked dot plots and report sheets, see below) to the students and aims at initiating mostly mathematizing activity regarding students' intuitive situative measures. It consists of several different phases, described below. In each phase, some design elements are utilized to implement the design principles.

The task, however, is not fully described through the design elements alone. Instead, the teaching-learning arrangement was constructed as to include the design experiment leader being present during the design experiments. Certain prompts were given by the design experiment leader, sometimes for every student pair, sometimes only when certain conditions were met. As these prompts had great impact on the learning processes, they need to be considered a part of the whole teaching-learning arrangement. Thus, the planned prompts for each

phase are also presented here, even if applied with a certain flexibility in order
to ensure adaptivity to students' processes.

Phase 1: Introducing the context

Task design puts the students into the role of consultants to Antarctic climate
researchers. The context is introduced to the students through a brief printed
description[1] (Fig. 7.1).

> The Norwegian research station *Troll Forskningsstasjon* (research station „Troll")
> in the Antarctic is used by researchers to investigate world climate. During the
> Antarctic winter (February to November), it houses up to 8 people. To be able to
> live and work under these extreme conditions, the researchers need very good
> preparation.

 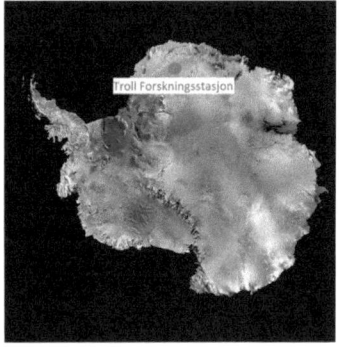

Fig. 7.1: The problem setting of the Typical Antarctic Temperatures Problem[2]

After reading the text, the students are given a distribution of daily temperatures
at the Troll research station for July 2002, along with another short description
(Fig. 7.2). Data was based on real observational data by Stroeve and Shuman
(2004).

---

1 Any design element presented here was translated from German in order to increase
   readability.
2 Troll research station image released under CC BY-SA 3.0 by Islarsh,
   https://commons.wikimedia.org/wiki/File:Troll_research_station_Antarctica.JPG; Antarctica image based on image released into public domain by Dave Pape,
   https://commons.wikimedia.org/wiki/File:Antarctica_6400px_from_Blue_Marble.jpg.

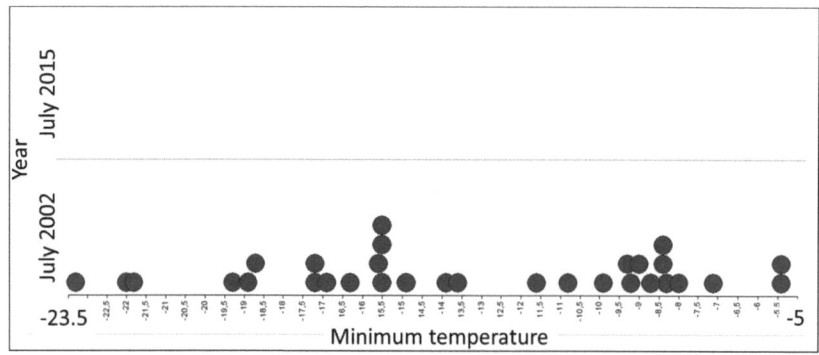

Fig. 7.2: Data for Phase 1 and 2 of the Typical Antarctic Temperatures Problem

This context implements design principle DP/Context, since weather was be-lieved to be a well-known phenomenon displaying short-term variability and long-term patterns. Antarctic weather, however, was also remote enough to the students' everyday experiences as to provide a phenomenon that still needed investigation. In this way, the Antarctic weather context was selected to enable and elicit mathematizing activities of structuring by providing some familiar but unusual phenomenon (see Table 7.1).

Tab. 7.1: Design elements of Phase 1 with corresponding design principles and intended effects

Design element	Design principle	Intended effects
Antarctic weather context	DP/Context	Identification of possible aspects of the phenomenon: Short-term variability, long-term patterns of center and spread, Possibility of predicting temperatures, typical temperatures, exceptional days
Single-year data	DP/Data	Identification of possible features of the data: Two peaks in data Range from -23.5 to -4.5 Middle 50% of data from -17 to -9
Stacked dot plot	DP/Technology	Using visual estimates for situative measures

The aim of this phase is to introduce the context and to guarantee sufficient understanding of phenomenon and data. Towards this end, the design experi-ment leader asks the students for their understanding of the situation and inter-venes when necessary (see Tab. 7.2 for prompts).

Tab. 7.2: Prompts for Phase 1 with conditions, as well as prompts to avoid

Condition	Prompt
Always	"What can you see in the diagram?"
Always	"There are 31 dots for the 31 days of July."
Students make no reference to frequencies	"There are three dots at -17, that means that there were three days with about -17°C."
Students show signs of interpreting x-axis as time	"You cannot see which dot is which day, you can only see that there were three days with -17°C in the whole of July."
Never	*Prompts explicitly referring aspects of phenomena like variability, typical temperatures, exceptional days or features of the data like mode, range, or middle 50%*
Basic understanding of plot reached	*Progression to Phase 2*

Another aim of Phase 1 is to introduce the students to the stacked dot plot, which they likely did not yet encounter at this point in their education. If students have trouble interpreting the diagram, prompts by the design experiment leader aim to ensure their general understanding of stacked dot plots by, for example, pointing out that there are 31 dots representing 31 days. Again, care should be taken not to pre-empt any structuring activity of the students. When a basic understanding of the plot is reached, the problem progresses to Phase 2.

Phase 2: Predicting phenomena

In this phase, the design experiment leader asks the students to give a prediction for ten days in July 2015 (which at the time of Cycle III still lay in the future). The students are encouraged to draw ten dots on the tablet screen. This step aims at creating the mathematizing situation of predicting phenomena. In this phase, the students can draw on their everyday experiences regarding variability of temperatures, thus prompting them to structure the phenomenon through situative mathematizing concepts by identifying aspects of phenomena. The limit to ten days is given so that the students do not simply copy the data from 2002, but rather have to identify aspects of the phenomenon that should also be reflected in data of 2015. This phase also serves as a way to access the students' initial reasoning so that their learning processes can be better evaluated later. After the students make their predictions, the problem progresses to Phase 3.

Tab. 7.3: Design elements of Phase 2 with corresponding design principles and intended effects

Design element	Design principle	Intended effects
Antarctic weather context	DP/Context	Identification of possible aspects of the phenomenon: Short-term variability, long-term patterns of center and spread, Possibility of predicting temperatures, typical temperatures, exceptional days
Single-year data	DP/Data	Identification of possible features of the data: Two modes Range from -23.5 to -4.5 Middle 50% from -17 to -9

Tab. 7.4: Prompts for Phase 2 with conditions

Condition	Prompt
Always	"Please make a prediction for ten days in July 2015 and draw them on the tablet screen."
Always	"Please explain your prediction to me."
Always	"Why this prediction and not any other?"
Students hesitate to commit to any prediction	"You can just give your best guess."
Students have explained their prediction	*Progression to Phase 3*

Phase 3: Dealing with year-to-year variability

After explaining their predictions, the students are given additional data from July 2003 and 2004 (Fig. 7.3).

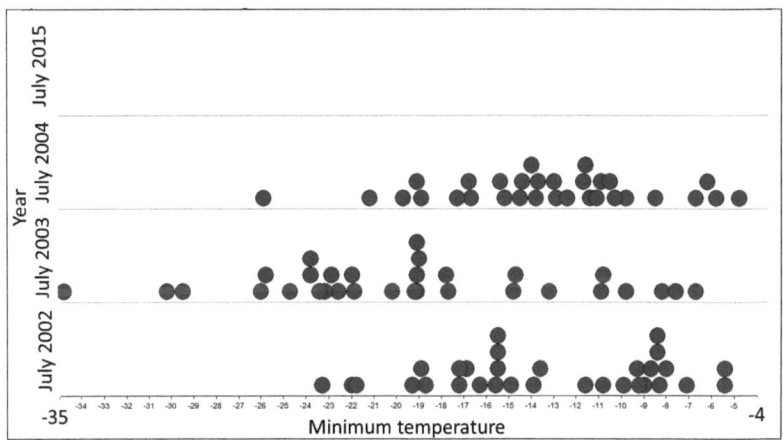

Fig. 7.3: Data for Phases 3 to 5 of the Typical Antarctic Temperatures Problem

The students then are asked whether they now needed to change their prediction and to do so accordingly. This again aims at prompting further structuring of the phenomenon according to their situative mathematizing concepts. The data are chosen in a way that might challenge predictions of 2015 that are based on the features of the modes of 2002, as 2003 shows features of the data that would not correspond to such aspects of the phenomenon (i.e. different spread and position of peak). The features of 2004 again implicate some stable aspects of Antarctic weather. Afterwards, the design experiment progresses to Phase 4.

Tab. 7.5: Design elements of Phase 3 with corresponding design principles and intended effects

Design element	Design principle	Intended effects
Three-year data	DP/Data	Identification of possible features of the data: Different modes Range from -35 to -4 Similar Middle 50% for 2002 and 2004
Antarctic weather context	DP/Context	Identification of possible aspects of the phenomenon: Short-term variability, long-term patterns of center and spread, Possibility of predicting temperatures, typical temperatures, exceptional days or years

Tab. 7.6: Prompts for Phase 3 with conditions

Condition	Prompt
Always	"Do you now need to change your prediction?"
Always	"Please explain your changes to me."
Students do not address aspect of variability of temperatures	"Another student once told me that you cannot really make a prediction, because it always changes. What do you think?"
Students do not address aspect of typical temperatures	"Another student once told me that it's enough to look at where the most dots are. What do you think?"
Students to not address aspect of exceptional days	"Another student once told me that you can ignore the day with -35°C because it only happened once. What do you think?"
Students refuse to commit to any prediction	"Do you think that it's simply impossible to predict the weather?"
Students have modified and explained their prediction	*Progression to Phase 4*

Phase 4: Reporting on Antarctic weather

In Phase 4, the students are given an empty *report sheet* (Fig. 7.4). Report sheets serve as the central design element for the whole design experiment series. They are introduced to the students as presenting the form of a brief report. Each report sheet consists of a sketch, some measures, and a summary. The sketch can be used to illustrate the structure of the phenomenon imposed by aspects of the phenomenon represented by measures and the summary gives an inference about the phenomenon based on the measures.

Report: Temperatures at *Troll Forskningsstasjon*	
Sketch	**Summary**
Minimum:	
Maximum:	
Typical:	

Fig. 7.4: Empty report sheet for the Typical Antarctic Temperatures Problem

The report sheets are given to the students without any explanations or require-ments other than to create a brief summary. In this way, the measures Minimum, Maximum, and Typical[3] provide situative measures, because the students are free to interpret them in an individual way: although minimum and maximum also constitute general measures, empirical results revealed individual interpre-tations of even these seemingly obvious general measures (e.g. 'reasonable' instead of absolute minimum and maximum)

When filling out a report sheet, the students need to engage in various math-ematizing activities. For providing a sketch and summary, they need to structure the phenomenon. The situative measures on the report sheets can be used to represent specific aspects of the phenomenon. And for finding actual values for the measures, they need to engage in formalizing. Thus, report sheets implement DP/Scaffold by providing a fixed form for measures the students need to fill out, prompting mathematizing activity.

---

3   Situative measures are indicated by use of upper-case letters, in order to distinguish be-tween situative measure ('Typical') and standard adjective ('typical').

Minimum and Maximum were chosen as well-known measures representing aspects of extreme temperatures, which were aspects of the phenomenon prominently identified in previous cycles of design research. A more prominent role is given to the situative measure Typical. 'Typical' provides an everyday idea commonly hoped to provide an interpretation for measures of center (Leavy & Middleton, 2011; Konold et al., 2002). Cycle II, however, revealed that it can also represent aspects of phenomena concerning variability (Büscher, 2017). In this case, Typical can act as a measure representing the aspect of Antarctic weather of a range of typical, likely temperatures that the students could have identified in their structuring activity of the earlier phases. Because the syllabus suggests that the students were only familiar with measures of center, Typical thus could provide a situative measure for an aspect of the phenomenon otherwise missing representing measures for the students. Typical is incorporated into the report sheets in order to ensure discussion and formalization of this promising situative measure.

This is a first step towards implementing DP/Formalizing, as the students need to commit to some values for Typical, allowing discussion about the correct values. The following empirical analysis (Section 7.2) reveals some of the variety of different formalizations observed in the design experiments. After the students create and explain their report sheet, the design experiment progresses to Phase 5.

Tab. 7.7: Design elements of Phase 4 with corresponding design principles and intended effects

Design element	Design principle	Intended effects
Empty report sheet	DP/Scaffold	Prompting students to provide an argument (summary) based on data as evidence (sketch and measures)
	DP/Formalizations	Prompting students to commit to concrete values for their situative measures
	DP/Measures	Use of Typical as a situative measure to represent aspects of the phenomenon identified earlier

Tab. 7.8: Prompts for Phase 4 with conditions

Condition	Prompt
Always	"Report sheets can be used to give a report to the researchers on the weather at Troll research station. Please give a report so that the researchers know for what to prepare themselves."
Always	"Please explain your report sheet to me."
Always	"Why did you choose your value for Typical?"
Students ask for the meaning of Minimum, Maximum, or Typical	"Just write down what you think fits best."
Students have created and explained their report sheet	*Progression to Phase 5*

Phase 5: Comparing report sheets

After creating their own report sheet, the students receive filled-in report sheets from other (fictitious) students (Fig. 7.5) and are asked to evaluate them. These filled-in report sheets confront the students with other interpretations of the situative measures and other structuring of the phenomenon. They provide conflicting summaries of the phenomenon; they use different types of sketches; and they give different values for the situative measures. Thus, although the phenomenon under investigation remains the same, as do the names of the situative measures, the inferences that can be drawn differ. This aims at creating the mathematizing situation of ensuring communication, as the students are able to observe that individual interpretations of situative measures hinders communication and should be avoided.

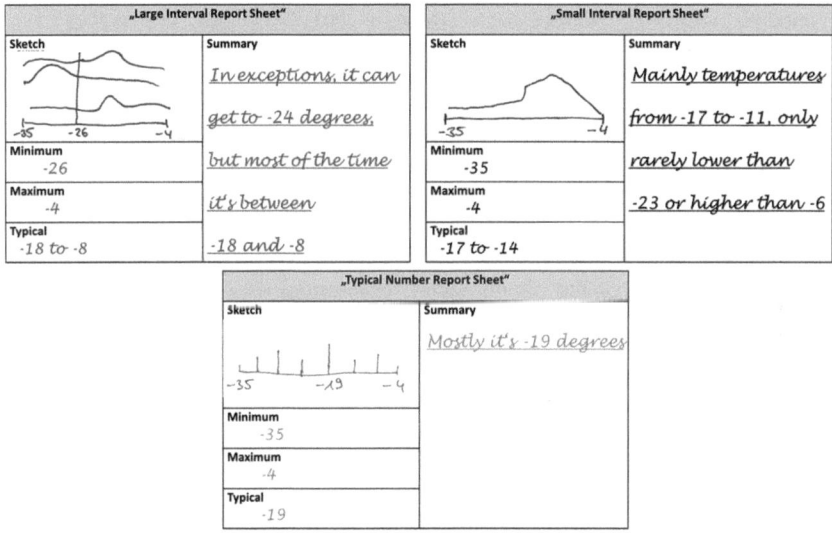

Fig. 7.5: Filled-in report sheets for Phase 5 of the Typical Antarctic Temperatures Problem

The filled-in report sheets are presented to the students without any further explanation of the situative measures involved. However, they were constructed with some aspects of the phenomenon and features of the data in mind, which could be identified by the students while investigating the filled-in report sheets (Tab. 7.9).

Tab. 7.9: A characterization of the situative measures on the filled-in report sheets

Measure	Formal characteristics	Features of the data	Aspects of the phenomenon
Minimum / Maximum [Large Interval Report Sheet]	Number	Main body of data	Reasonable possible temperatures
Minimum / Maximum [Small Interval Report Sheet]	Number	Boundary of data	Worst observed case, Exceptional days
Minimum / Maximum [Typical Number Report Sheet]	Number	Boundary of data	Worst observed case, Exceptional days
Typical [Large Interval Report Sheet]	Interval	Area encompassing most modes	Reasonably expected temperatures
Typical [Small Interval Report Sheet]	Interval	Highest density in 2002 and 2004	Most likely temperatures
Typical [Typical Number Report Sheet]	Number, single highest frequency	Mode of 2003	*No corresponding aspect*

In this activity, the students are asked which of the report sheets they preferred, and why. Prompts by the design experiment leader focus on getting the students to explain and to justify their decisions. This can cause various mathematizing activities: the students might engage in re-structuring the phenomenon to understand the filled-in report sheets, they might engage in representing to find differences in the aspects of the phenomenon represented by the different situative measures, and they might engage in formalizing to justify their own values. Finally, the students are asked to re-evaluate their own report sheet and whether they want to change it.

Tab. 7.10: Design elements of Phase 5 with corresponding design principles and intended effects

Design element	Design principle	Intended effects
Filled-in report sheets	DP/Formalizations	Eliciting formalizing activities because of differing values that cannot be easily understood; Introducing intervals as a valid possible formal characteristic of measures
	DP/Measures	Eliciting structuring and representing activities because of conflicting summaries and sketches; Introducing the idea of aggregating all available data
	DP/Language	Providing informal language means through informal language used in summaries

Tab. 7.11: Prompts for Phase 5 with conditions

Condition	Prompt
Always	"Which of the report sheets is right?"
Always	"What did you do different from the others?"
Always	"Do you now want to change your report sheet?"
Students' own report sheet contradicts a filled-in report sheet	"This report sheet says something completely different from yours. Why do you think is that so?"
Students do not acknowledge conflicting summaries	"The report sheets give very different summaries. Why do you think is that so?"
Students to not acknowledge conflicting formalizations	"One report sheet gives a single number for Typical, while the others use an area. What do you think is better?"

Two versions of the Typical Antarctic Temperatures Problem

One characteristic of design research is the possibility of flexibly adapting to challenges discovered during the design experiment series by adopting 'micro-cycles': controlled changes between design experiments within one cycle of design experiments (Gravemeijer & Cobb, 2006). For Cycle III, one micro-cycle concerns the data used in the Typical Antarctic Temperatures Problem. As shown in the analysis below, the large gap between 2004 and 2015 can provide an obstacle to students that, although important to solve eventually, distracted from the intended learning trajectory. This obstacle was observed in the design experiment with the first pair of students, Maria and Natalie. In order to eliminate this distraction, the labeling of the data was changed from '2002', '2003', and '2004' to '2012', '2013', and '2014'. The actual data were not affected by this change. In the analysis, reference to the Typical Antarctic Temperatures Problem of Cycle III without changes to the labeling are indicated by 'III-1a', references to the problem with changes to the labeling by 'III-1b'.

Concepts-in-action intended by design

As the goal of this thesis is to refine the design principles DP/Context, DP/Formalizations, and DP/Measures for teaching-learning arrangements focusing on mathematical literacy in statistics, the effects of the corresponding design elements for the students' learning processes need to be identified. This is done by comparing the students' concepts and activities to the intended effects of the design elements. These intended effects are articulated in terms of concepts-in-action that the various design elements intend to stimulate (Table 7.12). In the table, as well as in the analysis, concepts-in-action are coded as 'MCs/a' when concerning aspects of phenomena, 'MCs/f' for features of the data, 'MCs/c' for formal characteristics, and 'MCs/m' for measures. 'MCs' indicates a situative mathematizing concept, whereas 'MCg' indicates a general mathematizing concept (see Chapter 6 for a thorough description of the coding).

Table 7.12 presents an overview of the different concepts-in-action intended by design. The Antarctic weather context is intended to invoke some aspects of the phenomenon specified a priori, although in actual learning processes, the students identify their own aspects of the phenomenon. The table also shows how the increasing data available to the students might suggest focusing on different features of the data. Finally, it reveals the complexity of the design elements of the empty and filled-in report sheets, as they are intended to prompt the identification of several different measures, formal characteristics, features of the data, and aspects of the phenomenon.

The empirical analysis, however, also reveals some unexpected or unforeseen effects of the design elements, which can hinder as well as support the students' learning processes. This table helps in providing a list of possible concepts-in-action to be compared to the students' actual concepts-in-action.

Tab. 7.12: Intended concepts-in-action stimulated through the design elements of the Typical Antarctic Temperatures Problem.

Design element	Intended concepts-in-action
Antarctic weather context	||*MCs/a: short-term variability*|| ||*MCs/a: long-term patterns of center and spread*|| ||*MCs/a: possibility of predicting temperatures*|| ||*MCs/a: typical temperatures*|| ||*MCs/a: exceptional days*||
Single-year data	||*MCs/f: Two peaks*|| ||*MCs/f: Range from -23.5 to -4.5*|| ||*MCs/f: Middle 50% from -17 to -9*||
Three-year data	||*MCs/f: Different peaks*|| ||*MCs/f: Range from -35 to -4*|| ||*MCs/f: Similar Middle 50% for 2002 and 2004*||
Empty report sheet	||*MCg/m: Minimum*|| ||*MCg/m: Maximum*|| ||*MCs/m: Typical*|| ||*MCg/c: Method of calculation*|| ||*MCg/c: Interval*|| ||*MCs/a: typical temperatures*||
Filled-in report sheets	||*MCg/c: Interval*|| ||*MCg/c: Number*|| ||*MCs/c: Single highest frequency*|| ||*MCs/f: Main body of data*|| ||*MCs/f: Area encompassing most modes*|| ||*MCs/f: Highest density in 2002 and 2004*|| ||*MCg/f: Boundary of data*|| ||*MCg/f: Mode of 2003*|| ||*MCg/a: Worst observed case*|| ||*MCs/a: Exceptional days*||

‖*MCs/a: Most likely temperatures*‖
‖*MCs/a: Reasonable possible temperatures*‖
‖*MCs/a: Reasonably expected temperatures*‖

7.1.3 The Arctic Sea Ice Problem

In the Typical Antarctic Temperatures Problem, the students were introduced to report sheets and were familiarized with using measures to represent aspects of phenomena for reporting on phenomena. Chapter 4, however, described how learners need to broaden the contextual neighborhoods of their situated abstractions in order to develop general measures. To support such a process, the *Arctic Sea Ice Problem* dealt with a new phenomenon. Again, the problem progressed along several phases.

Phase 1: Introducing the context

This time, the students are put into the role of experts of Arctic sea ice tasked with giving a report on the state of Arctic sea ice. Again, the context is introduced to the students through a brief printed description (Fig. 7.6).

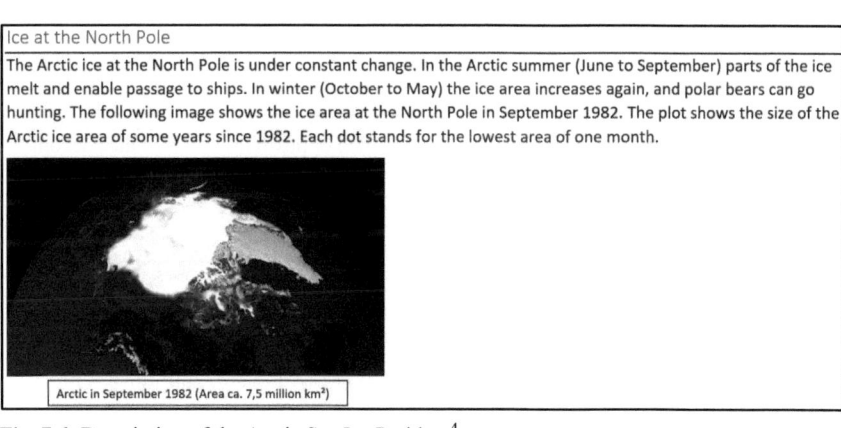

Fig. 7.6: Description of the Arctic Sea Ice Problem[4]

Along with the description they are given the data in Figure 7.7. The data used in the task are based on observational data from Fetterer, Knowles, Meier, and Savoie (2002). Again, a general understanding of the plots is ensured by the design experiment leader, giving great care not to pre-empt mathematizing activity by avoiding addressing possible aspects of the phenomenon.

---

4 Arctic sea ice image released into public domain by NASA/Goddard Space Flight Center Scientific Visualization Studio, https://svs.gsfc.nasa.gov/3445.

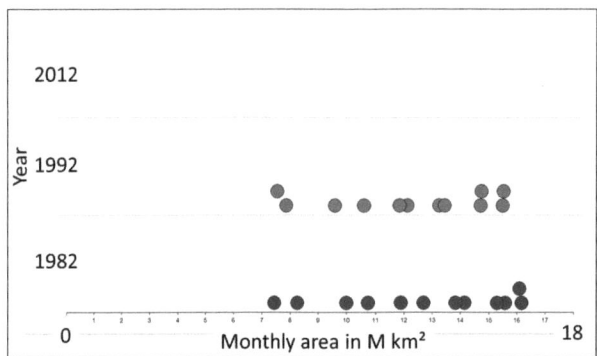

Fig. 7.7: Data for Phase 1 and 2 of the Arctic Sea Ice Problem

This context again implements the design principle DP/Context. Arctic sea ice, however, was expected to be less familiar to the students so that they would not have prior experience regarding aspects of the phenomenon, thus calling for more structuring activity. When sufficient understanding of the plots was achieved (e.g. that there were twelve dots for each year, representing the twelve months), the design experiment progressed to Phase 2.

Tab. 7.13: Design elements of Phase 1 with corresponding design principles and intended effects

Design element	Design principle	Intended effects
Arctic sea ice context	DP/Context	Identification of possible aspects of the phenomenon: Winter months, Summer months, Melting process,
Data from 1982 and 1992	DP/Data	Identification of possible features of the data: Modes Left-skewed Range from 7.5 to 16.5
Stacked dot plots	DP/Technology	Visual estimation of situative measures

Tab. 7.14: Prompts for Phase 1 with conditions and prompts to avoid

Condition	Prompt
Always	"What can you see in the diagram?"
Always	"There are 12 dots for the 12 months."
Students make no reference to frequencies	"There are two dots at 14 in 1982, that means that there were two months where the ice was 14 million km² at its lowest."
Students show signs of interpreting x-axis as time	"You cannot see which dot is which month, you can only see that there were two months with 14 million km²."
Never	*Prompts concerning aspects of phenomena like winter months, summer months, and or ice melt or features of the data like mode, skewness, or range*

Phase 2: Interpreting report sheets

The students are given three filled-in report sheets, which they are told give reports on Arctic sea ice in 2012 (Fig. 7.8). These data of 2012 are yet missing to the students, and they are asked to predict the data of 2012 based on the filled-in report sheets.

These report sheets differ from those employed in the Typical Antarctic Temperatures Problem by providing a blank entry for measures instead of always consisting of minimum, maximum, and Typical. This allows the filled-in report sheets to provide a variety of different measures that the students could use as inspiration for their own report sheet. In this way, the filled-in report sheets as well as the empty report sheets implement the design principle DP/Measures, as they allow the students to choose from a variety of possible different measures.

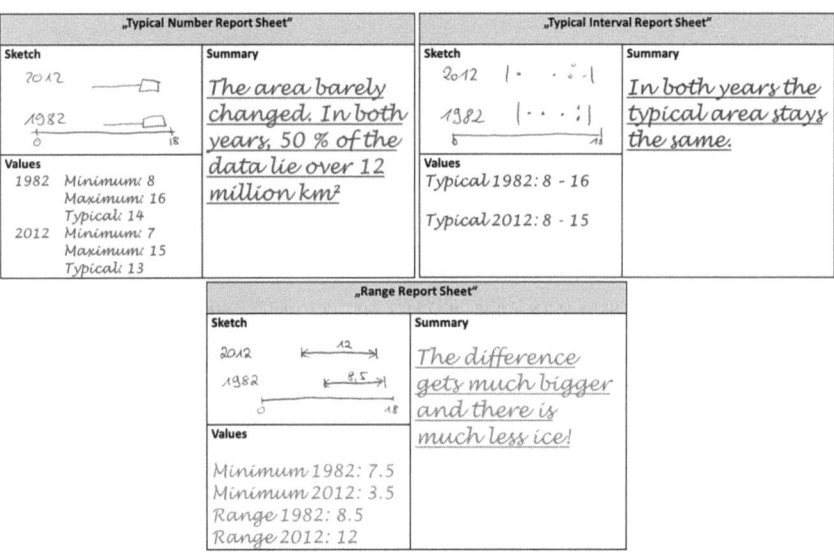

Fig. 7.8: Filled-in report sheets for Phases 2 and 3 of the Arctic Sea Ice Problem (titles not in original)

The filled-in report sheets are carefully constructed to show differences in formal characteristics, corresponding features of the data, and represented aspects of the phenomenon. In two of the report sheets, Typical takes different formal characteristics either as a number or an interval, calling for further activities of formalizing. Again, these filled-in report sheets are given to the students without further explanation of the measures involved, but several aspects of the phenomenon and features of the data could be identified by the students (Tab. 7.15).

Central to this phase is the observation that the report sheets contradict each other: the Typical Number Report Sheet and the Typical Interval Report Sheet both declare that there are no changes in Arctic sea ice, whereas the Range Report Sheet reports on dramatic changes. As the students do not have access to the data of 2012, the students are asked to hypothesize about possible reasons for these differences in perspective. After this, the design experiment progresses to the next phase.

Tab. 7.15: A characterization of the situative measures on the filled-in report sheets ('Typical Number Report Sheet = Typical Number Report Sheet, 'TIRS' = Typical Interval Report Sheet, 'RRS' = Range Report Sheet)

Measure	Formal characteristics	Features of the data	Aspects of the phenomenon
Typical [Typical Number Report Sheet]	Number	*No corresponding feature*	Beginning of winter
Typical [TIRS]	Interval	Main body of data	*No corresponding aspect*
Range [RRS]	Number Calculation through min. and max.	Spread of data	Severity of ice melt

Tab. 7.16: Design elements of Phase 2 with corresponding design principles and intended effects

Design element	Design principle	Intended effects
Filled-in report sheets	DP/Formalizations	Eliciting formalizing activities because of different formal characteristics of Typical in report sheets
	DP/Measures	Eliciting structuring and representing activities because of conflicting summaries and sketches; Introducing the idea of range as a measure spread
	DP/Language	Providing informal language means through informal language used in summaries

Tab. 7.17: Prompts for Phase 2 with conditions

Condition	Prompt
Always	"These report sheets report on the ice, but they already had access to data from 2012. Looking at the report sheets, what do you think the ice in 2012 was like?"
Students do not acknowledge conflicting summaries	"The report sheets give wildly different summaries. Why do you think is that so?"

Phase 3: Evaluating report sheets

In Phase 3, the students are given the missing data for 2012 (Fig. 7.9). The data show that the simple assertion that no change whatsoever has taken place does not hold. However, it was emphasized in Phase 2 that these were the data available to the creators of the filled-in report sheets. The students are asked whether looking at the data any changes in Arctic sea ice took place, and how they can be characterized. Afterwards, they are asked to evaluate the different report sheets and to judge whether any are right or wrong. They are also asked whether they now would find any reason for why the report sheets came to different conclusions.

The aim of this phase is to let the students perceive the influence of the choice of measure on the perspective on the phenomenon and the resulting inferences. As can now be seen, no report sheet simply made a wrong calculation or other formal errors. Both filled-in report sheets utilizing Typical – a measure that would have proven useful previously for reporting on Antarctic weather – assert no changes taking place. This is because such a measure represents the feature of the area of most dots. The corresponding aspect of the phenomenon, however, is simply not that important. For the phenomenon of Arctic sea ice, instead another aspect of the phenomenon is extremely relevant: how low the ice drops in summer. The Range Report Sheet utilizes measures that represent this aspect of Arctic sea ice, and thus can assert a change in the phenomenon. DP/Measures implemented in this conflicting way was meant to encourage the students to consciously reason about their choice of measure, initiating further processes of representing.

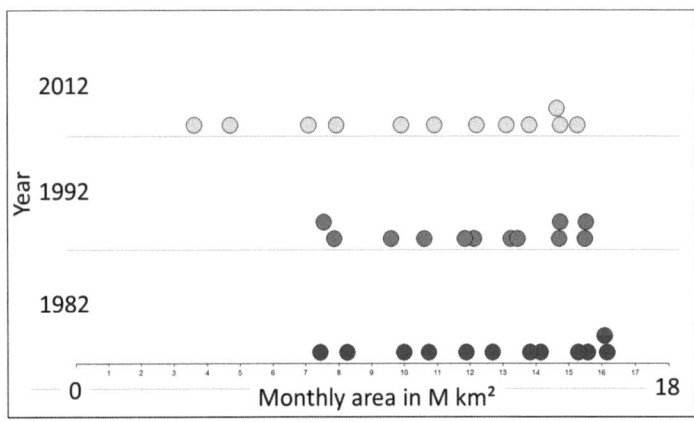

Fig. 7.9: Data for Phase 3 and 4 of the Arctic Sea Ice Problem

Tab. 7.18: Design elements of Phase 3 with corresponding design principles and intended effects

Design element	Design principle	Intended effects
Data of 1982, 1992, and 2012	DP/Data	Identification of possible features of the data: Increased spread in data
Arctic sea ice context	DP/Context	Need for finding measures to represent aspects of the phenomenon corresponding to features other than center
Typical on filled-in report sheets	DP/Context	Drawing connections between Antarctic temperatures and Arctic sea ice
Conflicting summaries on filled-in report sheets	DP/Measures	Noticing the influence of the chosen measure on the inferences to be made

Tab. 7.19: Prompts for Phase 3 with conditions

Condition	Prompt
Always	"Which of the report sheets is right?"
Students do not acknowledge conflicting summaries	"The report sheets give wildly different summaries. Why do you think is that so?"
Students to not acknowledge conflicting formalizations	"One report sheet gives a single number for Typical, while the others use an area. What do you think is better?"

Phase 4: Reporting on Arctic sea ice

After the students evaluate the filled-in report sheets, the students are asked to create their own report sheet. In this, they are free to adapt any of the filled-in report sheets or to choose or invent any measure of their liking. Afterwards, they are asked to explain their report sheet and to describe similarities and differences between their report sheet and the filled-in report sheets. The goal of this phase is not to reveal the students' own intuitive thinking, but the ways the task design influenced their thinking and how they adapted the ideas presented to them in the form of the filled-in report sheets.

Tab. 7.20: Design elements of Phase 4 with corresponding design principles and intended effects

Design element	Design principle	Intended effects
Empty report sheet	DP/Measures	Eliciting activities of representing and formalizing because of needed choice of measures

Tab. 7.21: Prompts for Phase 4 with conditions

Condition	Prompt
Always	"Please create your own report sheet for Arctic sea ice. You can do anything you want. You can use values from the report sheets, or invent your own."
Always	"What did you do different from the others?"
Always	"Which Values should one use to report on Arctic sea ice?"
Students do not use Typical in own report sheet	"Last time, you used Typical to report on Antarctic temperatures. Now you did not, why is that?"
Students use Typical in own report sheet	"Another student once told me that Typical was okay for temperatures, but not for ice. What do you think?"

Two versions of the Arctic Sea Ice Problem

The task design for the Arctic Sea Ice Problem was adapted in a micro cycle based on the experience gathered when observing the first pair of students, Maria and Natalie. The sequence as described above already refers to the modified task sequence III-2b. For the first pair of students in this design experiment series, Phase 2 was divided into two parts. In the first part, Maria and Natalie were given only the two report sheets utilizing Typical. Only after they had interpreted both report sheets, were they given the conflicting Range Report Sheet. The reason for this was to ensure that they actually noticed the conflicting summaries. This stepwise interpretation of the report sheets, however, proved artificial, and without the conflict introduced through the Range Report Sheet, only little activity took place during the first part. Thus, for subsequent student pairs (III-2a), all three filled-in report sheets were handed out at once (III-2b), and the design experiment leader prompted the students to explain the different summaries whenever they failed to notice.

Concepts-in-action intended by design

As with the Typical Antarctic Temperatures Problem, the design elements of the Arctic Sea Ice Problem also intend to stimulate certain concepts-in-action. Again, Table 7.21 shows different possible aspects of the phenomenon of Arctic sea ice as well as features of the data that could be identified by the students. The design element of the three-year data from 1982, 1992, and 2012 is intended to elicit concepts-in-actions concerning features of spread and aspects of ice melt. The filled-in report sheets, however, do not all correspond to these features: although the measure range would address such an aspect, the measure Typical would conform more to a feature of the main body of data, thus obscuring the more important ice melt. The task design aims at initiating discussion between these possible focal points of different measures, formal characteristics, features of the data, and aspects of the phenomenon. This list of intended concepts-in-action then enables a comparison of the intended versus the actual students' concepts-in-action from the learning processes.

Tab. 7.22: Intended concepts-in-action stimulated through the design elements of the Arctic Sea Ice Problem.

Design element	Intended concepts-in-action
Arctic sea ice context	||MCs/a: winter months||
	||MCs/a: summer months||
	||MCs/a: melting process||
Two-year data	||MCs/f: modes||
	||MCs/f: left-skewed||
	||MCs/f: spread from 7.5 to 16.5||
Three-year data	||MCs/f: increased spread in data||
	||MCs/a: increasing severity of ice melt||
Empty report sheet	||MCg/m: range||
Filled-in report sheets	||MCg/m: range||
	||MCs/m: Typical||
	||MCg/c: number||
	||MCg/c: interval||
	||MCg/c: calculation through min. and max.||
	||MCs/f: main body of data||
	||MCg/f: spread of data||
	||MCs/a: beginning of winter||
	||MCs/a: severity of ice melt||

7.2 Empirical reconstruction of students developing Typical

This part of the analysis follows the learning processes of two pairs of students from Cycle III, Maria & Natalie and Quanna & Rebecca. The analysis starts with Phase 2 of the Typical Antarctic Temperatures problem and presents a longitudinal reconstruction of both pairs' learning processes.

The focus of this analysis lies on the mathematizing side and follows the students' development of the situative measure Typical. This analysis identifies students' situative mathematizing concepts and possible pathways towards general mathematizing concepts through students' mathematizing activities. Thus, the main focus of this analysis lies on research question RQ1, concerning the development of measures. However, because the students' reflective activities are also taken into account, some insights also are provided concerning RQ2 (the specification of reflective concepts) and RQ3 (the interplay between mathematizing and reflective activities).

The development is tracked by identifying the students' mathematizing concepts-in-action and the mathematizing activities encoded in their theorems-in-action. The mathematizing concepts-in-action are then assigned to the categories of *measures, formal characteristics, features of data,* and *aspects of phenomena.* The development of Typical then is reconstructed by a growing complexity of concepts-in-action and increasingly sophisticated mathematizing activities.

The analysis also provides insights into how students' reflective activities and concepts also play an important role in their mathematizing activities. Following the method outlined in Chapter 6, the analysis begins with identifying students' mathematizing activities and concepts. In a second step, the students' reflective activities are identified based on their reflective theorems-in-action. Following that, the situative and general reflective concepts-in-action are reconstructed based on those activities.

This analysis expands on two articles in which preliminary and shorter versions have been presented (Büscher, 2018; Büscher, 2017).

7.2.1 Maria and Natalie: Developing with a focus on structuring

The analysis first follows the learning processes of Maria and Natalie. Both students are highly motivated and do not hesitate to present their spontaneous and unfinished ideas. Their learning process is characterized by an intense interaction, visible for example by the fact that they often finish each other's sentences.

Typical Antarctic Temperatures Problem – Episode 1

This analysis starts with Phase 3 where the mathematizing activities of the students increase dramatically. The students had been introduced to the context of the problem during Phase 1. In Phase 2, they provided a prediction for ten days in July 2015, which was mostly based on the feature of the modal clump in 2002. Immediately after the students receive the additional data of 2003 and 2004, and even before the design experiment leader asks them to give a prediction, they begin to discuss the data. The discussion goes on for two minutes, after which the design experiment leader intervenes.

III-1a-MN; Phase 3; Start: 21:00
Previously, the students discussed the data of 2002, 2003, and 2004 in order to find a prediction based on these years. Now, the design experiment leader intervenes and asks the students to explain their thoughts. Maria and Natalie observe patterns in the data, but struggle to find their meaning.

| 309 | DL | I don't quite get it. | Jetzt komm ich da grad nicht ganz so hinterher. |
| 310 | | *[both laugh]* | *[beide lachen]* |

311	DL	What do you say, what, uuh #what – yes exactly.	Was sagt, was ehm, #was - ja genau.
312	M	#We're just thinking what the relationship, like, how you'd...	#wir überlegen gerade, welchen Zusammenhang, also, wie man das...
313	N	Well, because we just want to know what changes between each year. And then we said that there [points to 2003], it, like, came apart.	Ja, weil wir wollen halt wissen, was sich da zwischen einem Jahr halt immer ändert. Und dann haben wir halt gesagt, dass es da [zeigt auf 2003] halt so auseinandergegangen ist.
314	M	And here [points to 2004] it's again as before.	Und da [zeigt auf 2004] wieder wie beim alten.
315	N	And here again it's...	Da dann wieder ...
316	M	Well not #exactly but...	Ja nicht #ganz aber...
317	N	#downwards. Except it's somehow exactly the wrong way around. #You know, compared to this [points to 2002].	#runter. Bloß, dass es halt genau irgendwie so falschrum ist. #Find ich, als das hier [zeigt auf 2002].
318	M	#Yes.	#Ja.
319	DL	Um.	Mhm.
320	M	Well, you know, basically this [points to 2004] is like similar to that [points to 2002], but this [points to 2003] is just different. And then somehow you can't see it again.	Ja, ich find ja, das [zeigt auf 2004] ist ungefähr das ähnliche wie das [zeigt auf 2002], aber das [zeigt auf 2003] ist halt anders. Und dann kann man das irgendwie wieder nicht sehn.
321	N	So, there [points to 2004, around -12 °C] are, like, like most of the dots, and here [points to 2002, -12 °C] are hardly any. And there [points to 2002, -8 °C] are the most and here [points to 2004, -8 °C] are hardly any. So (3s) but you can't just say that there always is a pattern. Right?	Also da [zeigt auf 2004, bei -12°C] sind halt, halt die meisten Punkte und hier [zeigt auf 2002, -12°C] sind fast gar keine. Und da [zeigt auf 2002, -8°C] sind die meisten und da [zeigt auf 2004, bei -8] sind fast gar keine. Also (3s) aber man kann ja auch nicht sagen, dass immer ein Muster ist. Ne?

Analysis of mathematizing side. In this scene, the pair encounters the mathematizing situation of predicting phenomena. The students try to find a way to characterize "the relationship" (#312) of "what changes between each year" (#313). This shows how the students perceive the phenomenon by a theorem-in-action. Theorems-in-action are propositions held to be true by the learners and are connected to their mathematizing activities. In this case, the students identify a new aspect of the phenomenon, and thus engage in the mathematizing activity of structuring (marked as <MA/s>) the phenomenon: <MA/s: Antarctic temperatures comprise a change from year to year>.

Theorems-in-action describe relations between concepts-in-action, which are the categories held relevant by the learners. From this structuring theorem-in-action, the corresponding concept-in-action can be inferred which guides the students' perception in this moment: the situative mathematizing concept-in.action of the aspect of the phenomenon (marked by 'MCs/a') of the $||MCs/a: year-to-year change||$.

They also observe several features of the data: In 2003, the students hold that the data "like, came apart" (#313). This feature of the data (marked by 'MCs/f') can be used to reconstruct an additional situative concept-in-action of $||MCs/f: coming-apart||$. This is a situative concept-in-action, because it is bound to the specific data at hand and does not conform to any concept of formal statistics. However, the students use it to indicate that the data are more 'densely packed' in one area than in the others. Thus, it could be a situative precursor to a general concept describing a feature of the data like $||MCg/f: density||$.

In identifying additional features of the data, the students find that the data of 2002 are similar to 2004, albeit "somehow exactly the wrong way around" (#317). This feature of the data the students focus on here is reconstructed as the students' situative concept-in-action of $||MCs/f: wrong way around||$. This could be interpreted as a precursor to the general concept of $||MCg/f: skewness||$ of the data, because both concepts refer to the orientation of the data. Natalie explains this situative concept by tracking the location of the $||MCs/f: area of most dots||$: whereas in 2004, the most dots are at -12°C and almost none at -8°C, the order is switched for 2002 (#321). This is reconstructed as the situative concept-in-action of $||MCs/f: switching pattern||$. This feature of the data, however, seems to be in conflict with their contextual knowledge, as $<MA/s: a switching pattern cannot correspond to an aspect of Antarctic temperatures>$ ("you can't just say there always is a pattern", #321).

The students do not seem to come to a conclusion regarding the $||MCs/a: year-to-year change||$. Their structuring activity seems to conclude without a definite result, implying their theorem-in-action $<MA/s: None of the features of the data correspond to the year-to-year change>$.

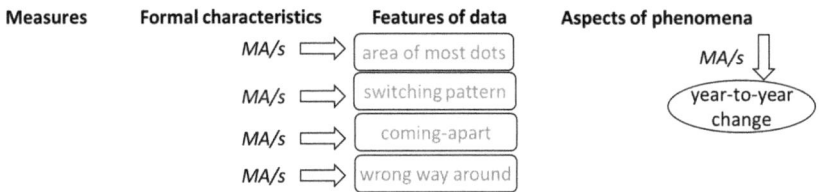

Fig. 7.10: Maria and Natalie's identify several features of the data and an aspect of the phenomenon, but do not find connections between them

Analysis of reflective activities. So far, no reflective activity takes place, as there is no reflective situation.

Analysis of reflective concepts. So far, no reflective concepts can be identified. In the following, the analysis of the reflective activities and concepts are only explicated when reflective activities take place.

Typical Antarctic Temperatures Problem – Episode 2

The next Episode takes place few minutes later, as the students try to give a prediction for 2015. In order to do so, they try to find a way to describe the ||*MCs/a: year-to-year change*||.

III-1a-MN; Phase 3; Start: 22:30
Previously, the students showed difficulties in describing the change from year to year. Now, Maria and Natalie try to give a prediction for ten days in July 2015.

346	N	From 2004 to 2015 there are... #eleven years... So, eleven. This *[points to 2015]* then ought to be roughly like that one *[points to 2003]*. Right?	Von 2004 bis 2015, das sind... #elf Jahre... Und elf. Das *[zeigt auf 2015]* müsste dann ungefähr sein wie das *[zeigt auf 2003]*. Oder?
347	M	#But, uh, one ought to first take a look how many degrees it increased or decreased. In general. In two years.	#Aber ehm, man müsste doch erstmal gucken, wie viel Grad das gestiegen oder gesunken ist. Generell. In zwei Jahren.
348	M	Nah. We could just take a look, well no, we're not supposed to calculate *[laughs]*.	Nee. Wir können doch erst mal gucken, ach nee, wir sollen ja nicht rechnen *[lacht]*.
349	DL	You could calculate, #if you want to.	Ihr könnt auch rechnen, #wenn ihr wollt.
350	M	#Okay.	#Achso
351	DL	You just don't have to.	Ihr müsst nur nicht.
352	M	We could just take a look at how many degrees, like the diff-... like, we could...	Wir können noch gucken, wie viel Grad, also der Unt-... also wir kön-nen...
353	N	You mean like #average, how...	Du meinst #Durchschnitt, was...
354	M	#the average and then we take a look how the average changed in two years.	#Den Durchschnitt und dann gucken, wie der sich der Durch-schnitt in zwei Jahren verändert hat.

Analysis of mathematizing side. This Episode provides an example for both mathematizing situations, predicting phenomena (because the students are try-ing to find a valid prediction) and ensuring communication (because the stu-dents are concerned with finding an objective prediction). Natalie tries to give a prediction by identifying a feature the data of 2015 should have. Following the ||*MCs/f: switching pattern*|| reconstructed earlier, she notes that "This *[points to 2015]* then ought to be roughly like that one *[points to 2003]*" (#346). In this way, Natalie engages in structuring by predicting features of the data that 2015 should exhibit, here reconstructed as the theorem-in-action <*MA/s: the data of 2015 should exhibit similar features to the data of 2003*>.

At this point, Maria disengages with the idea of following an ||*MCs/f: swtiching pattern*|| by suggesting to first look at "how many degrees it increased or decreased. In general" (#347). In this new idea, Maria is trying to summarize each year under a single value, in order to calculate the difference between 2002 and 2004. This single summary value is not a feature of the data. Instead, in this analysis it is reconstructed as the aspect of Antarctic temperatures of an abstract ||*MCs/a: general temperature*||, which gets used in Maria's structuring activity <*MA/s: the phenomenon of Antarctic temperatures comprises a general temperature*>. The aspect of ||*MCs/a: general temperature*|| can then be used to find the ||*MCs/a: year-to-year change*|| (#347, how the temperatures "generally" rise or fall).

This ||*MCs/a: general temperature*|| marks a first aspect of a phenomenon that was not anticipated to be identified by design. The reason might be that within everyday contexts it is common to summarize varying temperatures under a general temperature, often represented by the average. What she exactly means by this ||*MCs/a: general temperature*||, however, remains unclear.

Maria also introduces a measure she already knows, the ||*MCg/m: average*||. She introduces this measure while trying to find a way to describe the ||*MCs/a: year-to-year change*||. This act of finding a fitting description show her mathematizing activity of representing ('MA/r') belonging to the reconstructed theorem-in-action <*MA/r: differences in averages represent the year-to-year change*> (you have to "take a look how the average changed", #354). It is interesting to note that although the students previously referred to the ||*MCs/f: area of most dots*||, a mathematizing concept which could be a predecessor to the general concept of ||*MCg/f: center*||, they do not relate the ||*MCs/f: area of most dots*|| to the ||*MCg/m: average*||, which is a measure with the corresponding feature of the data of ||*MCg/f: center*||. This could be a sign that, although they readily make use of the average, their individual knowledge of the average could also develop throughout this design experiment.

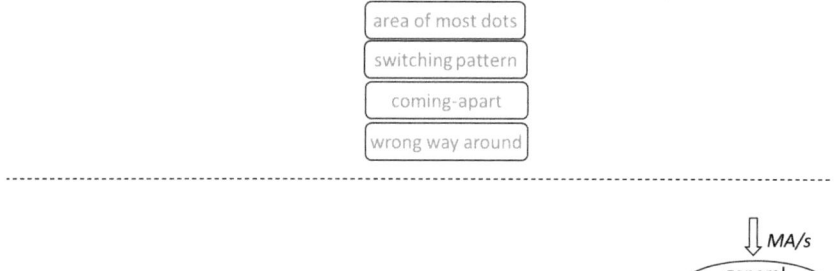

Fig. 7.11: Maria and Natalie identify two new aspects of the phenomenon. They engage in representing by connecting these aspects to the average. The features of the data identified earlier fade from discussion

Typical Antarctic Temperatures Problem – Episode 3

Following this scene, the students try to utilize the average for their prediction.

III-1a-MN; Phase 3; Start: 23:30
Previously, Maria and Natalie used differences in averages to represent the year-to-year change. Now, they try to use the average for their prediction. This leads them to assert a cooling of temperatures in a linear way, which clashes with their context knowledge.

377	M	No. We… I wanted to know how much the average changed between these two years. #So that you…	Nein. Wir… Ich wollt wissen, wie viel sich der Durchschnitt geändert hat zwischen den zwei Jahren hier. #Das man das…
378	N	#Yes, from -12 to…	#Ja von -12 auf…
379	M	So, 2… 2 degrees. Basically. So, got colder again. That's 2 degrees. By 2004. Right? Because like there are…	Also 2… 2 Grad. Sozusagen. Also ist wieder kälter geworden. Das sind 2 Grad. Bis 2004. Oder? Weil da sind da doch…
380	N	2 degrees colder.	2 Grad kälter.
381	M	Yes. Nah, warm… #Yeah, colder.	Ja. Nee, wä… #Ja kälter.
382	N	#And then in…	#Und dann in…
383	M	Nah. What? Ah, okay, 2 #degrees colder.	Nee. Hä? Ja doch 2 Grad #kälter.
384	N	#Wait! If in t… two years it gets 2 degrees colder, then every year it gets 1 degree colder, so you have to…	#Warte. Wenn das bei zw… zwei Jahren, 2 Grad kälter wird, dann wird's ja in jedem Jahr ein Grad kälter also muss man…
385	M	Nah. Oh, right!	Nee, ja doch!
386	N	13 degrees colder average temperature. Right?	13 Grad kälter, die Durchschnittstemperatur. Oder?

387	M	Yes.	Ja.
388	N	But that's too much. Right?	Das ist aber zu viel. Oder?
[...]			
395	M	But you still need to take a look, because over the course of the years it's also, like, #many more cars and nuclear reactors and such. And... *[laughs] (incomprehensible)*	Aber man muss dann ja noch gucken, weil in den laufen des Jahres ist auch doch #viel mehr Autos und Atomwerke und so. Und... *[lacht] (unverständlich)*
396	N	#that changes. *[laughs] (incomprehensible)*	#das ändert sich. *[lacht] (unverständlich)*
397	M	No. It is said that it gets warmer. So, it somehow needs to get warmer. And not colder. *(3s)* Right?	Nein. Man sagt ja, dass es wärmer wird. Also muss es ja in irgendeiner Weise wärmer werden. Und nicht kälter. *(3s)* Oder?

Analysis of mathematizing side. Now that the students have found a measure that represents the $||MCs/a:$ *general temperature*$||$, they use it to quantify the $||MCs/a:$ *year-to-year change*$||$. They argue that, since in the span of two years from 2002 and 2004 the average temperature dropped by an estimated 2 degrees (#379, not calculated, but visually estimated), in a span of 13 years it would drop by 13 degrees (#386). Thus, based on their use of the average in representing the $||MCs/a:$ *year-to-year change*$||$, they identify a new aspect of Antarctic temperatures: $<MA/s:$ *The phenomenon of Antarctic temperatures comprises a linear cooling>* of 1 degree per year.

This, however, irritates the students ("but that's too much", #388), as based on their contextual knowledge of the phenomenon, they also identify the aspect of $||MCs/a:$ *global warming*$||$ ("it needs to get warmer", #397). Again, their activity of structuring phenomena gets disturbed, as they can find no features of the data corresponding to the aspect of $||MCs/a:$ *global warming*$||$, and this aspect of the phenomenon even contradicts the aspect of $||MCs/a:$ *linear cooling*$||$ ("it somehow needs to get warmer. And not colder", #397).

In this way, they can productively use their knowledge of the phenomenon to inform their mathematizing activities. Although the students show a very individual understanding of global warming (13 years being too short a timeframe for noticing significant differences in average temperatures) which they use to structure the phenomenon (an effect not anticipated, as in this case they do not base their predictions on data), this aspect helps them in validating their inference of a $||MCs:$ *linear cooling*$||$. Temperatures simply do not behave in such a way as to linearly decrease each year. Thus, the average might not be such a good measure to use for a prediction, at least not applied in this way.

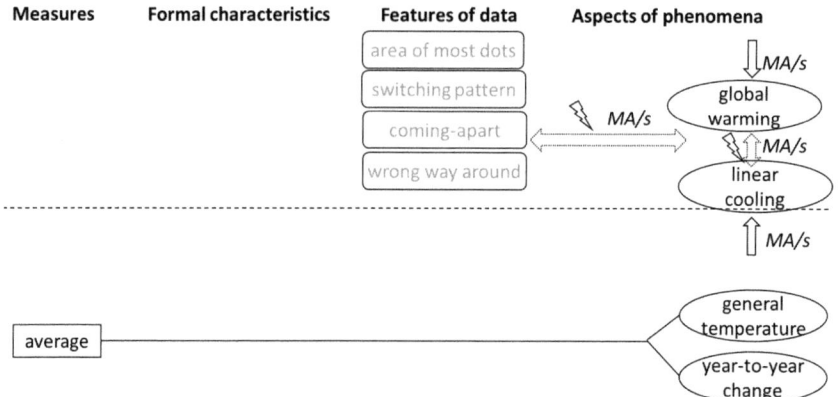

Fig. 7.12: Maria and Natalie engage in structuring and identify two conflicting aspects of the phenomenon. No feature of the data corresponds to their aspects of the phenomenon

Typical Antarctic Temperatures Problem – Episode 4

A minute later, the design experiment leader prompts the students to explain their use of the average for the prediction.

III-1a-MN; Phase 3; Start: 25:20
Previously, Maria and Natalie used the average to find the general temperature of 2015 to inform their prediction. Now, the design experiment leader asks them to explain their reasoning. The students try to find the meaning of average.

408	DL	What, what do you say as to why the average is so good? What does it tell you?	Was, was sagt ihr, warum ist der Durchschnitt denn eigentlich so gut? Was sagt der einem denn?
409	M	Well, that it gets colder. Like, that the climate is...	Ja, dass es kälter wird. Also, dass das Klima sich...
410	N	*(3s)* It's just more exact, I believe.	*(3s)* Das ist einfach genauer, glaube ich.
411	M	What?	Wie?
412	N	This now doesn't tell you anything about every single day, but #like how it was overall.	Das sagt ja jetzt nichts über jeden einzelnen Tag aus, aber #so, wie es insgesamt so war.
413	M	*#(incomprehensible)*	*#(unverständlich)*
414	DL	Um.	Mhm.
415	M	Well. But, what, like, they say it's getting warmer.	Ja. Aber, was, man sagt ja, dass es jetzt wärmer wird.
416	N	Yes, they say that, Maria.	Ja, sagt man, Maria.
417	M	Yeah, they say. As if they'd say that without a reason.	Ja sagt man. Als ob die das ohne Grund sagen würden.
418	N	Well, but somehow we then calculated wrongly.	Ja aber irgendwie haben wir dann falsch gerechnet.

Analysis of mathematizing side. Maria starts to explain the usefulness of the average by repeating that <*MA/r: the difference of averages represents the linear cooling*|| of Antarctic temperatures (#409). She then gets interrupted by Natalie, who distinguishes between "every single day" and "how it was overall" (#412). It is unclear what exactly she means by this. Because she relates "how it was overall" to the average, this can be interpreted as Natalie again drawing on the ||*MCs/a: general temperature*|| represented by the average. The newly identified aspect of ||*MCs/a: individual days*|| remains without explanation for now. The identification of this new aspect, however, is a sign of structuring activity, reconstructed through the theorem-in-action <*MA/s: The phenomenon of Antarctic temperatures comprises individual days and general temperatures*>. Thus, the students' view on the phenomenon becomes more pronounced.

This structuring, however, does not resolve the conflict regarding the conflicting aspects of the phenomenon of ||*MCs/a: global warming*|| and ||*MCs/a: linear cooling*||: the students still hold that "it's getting warmer" (#415), implicitly holding that <*MA/s: the phenomenon of Antarctic temperatures cannot comprise both, global warming and linear cooling*>. The authority of the contextual knowledge concerning ||*MCs/a: global warming*|| in fact is so strong as to induce doubts about their methods of calculation ("somehow we then calculated wrongly", #418).

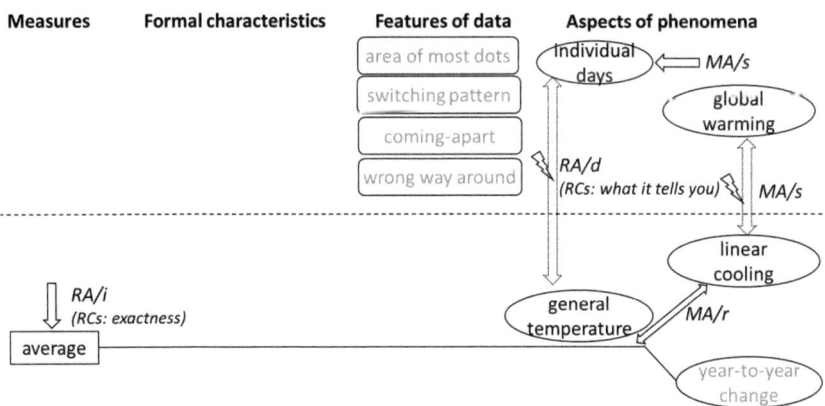

Fig. 7.13: Maria and Natalie identify a new aspect of the phenomenon. They contrast the different aspects of the phenomenon in structuring activity, which is also aided by reflective activities

Analysis of reflective activities. This Episode is an example of the reflective situation of rejecting measures, as the students find justifications for rejecting the average. Natalie explains the average not by relating it to phenomenon or data, but she provides an intuitive meaning of being "simply more exact"

(#410). This explanation can be interpreted as an instance of the reflective activity of identifying patterns of thought ('RA/I'), reconstructed as the reflective theorem-in-action $<RA/i$: the average is more exact$>$. In explaining what she means by that, she defines a case in which the average would not be useful: the average does not "tell" about $||MCs/a$: individual days$||$ (#412). Thus, she does not engage in representing, but does the opposite. She contrasts the aspects of $||MCs/a$: individual days$||$ and $||MCs/a$: general temperature$||$ and finds an aspect of the phenomenon that is not represented by the average. This can be interpreted as an instance of the reflective activity of denominating risks and limits ('RA/d') with the corresponding reflective theorem-in-action $<RA/d$: the average does not tell you anything about individual days$>$.

Analysis of reflective concepts. Based on these reflective theorems-in-action, the corresponding reflective concepts-in-action can be identified. For identifying patterns of thought, Natalie draws on her situative reflective concept-in-action ('RCs') of $||RCs$: exactness$||$. It is not entirely clear what she means by that. Still, this reflective concept-in-action seems to be important to her mathematizing activities, since it informs her use of the $||MCs/g$: average$||$.

In her theorem-in-action underlying her reflective activity of denominating risks and limits, Natalie draws on the situative reflective concept of $||RCs$: what it tells you$||$. This reflective concept-in-action was introduced by the design experiment leader (#408). Natalie is able to adopt it to differentiate measures in terms of which aspects they represent and what could be understood of the phenomenon based on those measures: the average 'tells you' nothing about $||MCs/a$: individual days$||$, and if you only used the average, you wouldn't know anything about $||MCs/a$: individual days$||$. Thus, $||RCs$: what it says$||$ could be a situative precursor to the general reflective concept ('RCg') of $||RCg$: perspectivity$||$, the concept concerning the fact that use of any mathematics can determine the view on phenomena, for better or worse.

Typical Antarctic Temperatures Problem – Episode 5

Shortly after, the design experiment proceeds to Phase 4, and the students work on their own report sheet (Fig. 7.14). The students' mathematizing activities, however, lack in explicitness, as they mostly agree with each other, and if not, seem to reach agreements without explicating their reasons. The design experiment leader, however, also does not intervene, so as not to disturb their creation of their own report sheet. Thus, Phase 4 of the design experiment is not part of this analysis.

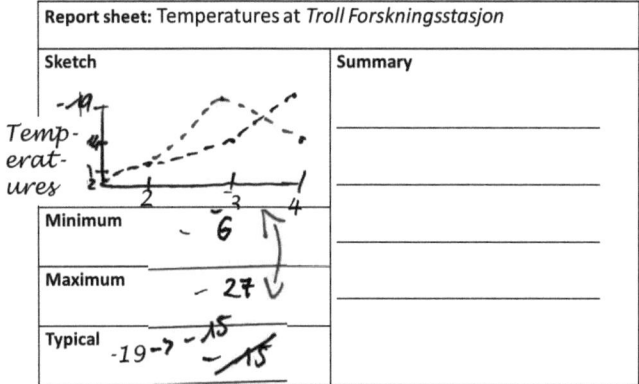

Fig. 7.14: Maria and Natalie's own report sheet for the Typical Antarctic Temperatures Problem

Because the students take a long time in creating their own report sheet without much explanation, the design experiment leader finally intervenes before the students produce a summary for the report sheet. He prompts the students to investigate the filled-in report sheets, thus progressing the design experiment to Phase 5. In this phase, the students are confronted with the three filled-in report sheets featuring different formalizations of the situative measure Typical. This phase intends to elicit discussion about this situative measure, which in the design experiment of Maria and Natalie, so far did not have an important role. The analysis resumes at this point.

III-1a-MN; Phase 5; Start: 42:30
Previously, Maria and Natalie created their own report sheet and the design experiment progressed to Phase 5. Now, the students read the filled-in report sheets and discover the possibility of using an interval as a formal characteristic of Typical.

733	N	Typical. They have... So, we have -15, *[reading report sheets]* -19, -18 to -8?	Typisch. Die haben... Also wir haben -15, *[liest Steckbriefe]* -19. -18 bis -8?
734	M	Uh, you can't really ...	Hä, das kann man ja nicht...
735	N	Okay, but that is actually better, compared to indicating just one, because it is always changing. Although most of the time it's between two temperatures.	Naja, aber das ist eigentlich besser, als wenn man nur eine angibt, weil es sich ja auch immer wechselt. Aber dass es meistens zwischen zwei Temperaturen ist.
736	M	Wait, where?	Warte wo?
737	DL	Could you somehow explain this using an example? Or, what do you mean? It's better? That it's changed? What do you mean?	Könntet ihr das mal im Beispiel irgendwie ein bisschen sagen. Oder wie meinst du das? Das ist besser? Dass man mal das wechselt. Wie meinst du das?
738	N	Well, it's a bit easier to understand, you know, if you, uh, if you like	Also, das ist einfach ein bisschen verständlicher find ich, wenn man,

indicate two things like -18 to -8, instead of indicating just one thing, because this is actually an average temperature and I believe that you can understand this better, because it changes from day to day by one or two temperatures.	ähm, halt zwei Sachen angibt halt so - 18 bis -8, als wenn man jetzt nur eine Sache angibt, weil das ist ja wirklich eine Durchschnittstemperatur und das kann man glaube ich einfach besser verstehen, weil es wechselt sich ja auch von Tag zu Tag mal um ein zwei Temperaturen.

Analysis of mathematizing side. On their own report sheet, the students gave the single number '-19' as Typical, whereas two of the filled-in report sheets use an interval. In this scene, Maria and Natalie discuss whether Typical should take the form of an interval or a single number ("Uh, you can't really" take an interval, #734, but "but that is actually better, compared to indicating just one", #735). This marks the first occurrence where the students do not refer to measures, features of the data, or aspects of the phenomenon. Instead, they refer to possible formal characteristics ('MCg/c') of Typical as $||MCg/c: number||$ or $||MCg/c: interval||$. The identification of a formal characteristic of a given measure then is an act of the mathematizing activity of formalizing ('MA/f') with the underlying theorem-in-action $<MA/f: Typical\ can\ either\ take\ the\ form\ of\ an\ interval\ or\ a\ number.$

For deciding the fitting formal characteristic, Natalie turns towards structuring the phenomenon by identifying another aspect: $<MA/s: the\ phenomenon\ of\ Antarctic\ temperatures\ comprises\ always\ changing\ temperatures>$ it is always changing. Although most of the time it's between two temperatures", #735). This new aspect of $||MCs/a: always\ changing||$ then provides the justification for choosing an $||MCg/c: interval||$ over a $||MCg/c: number||$: she states that "It changes from day to day" and so it's better to "indicate two things" (#738). This 'indicating two things' can be interpreted as a preference to the $||MCg/c: interval||$, and thus an instance of the mathematizing activity of formalizing that $<MA/f: Typical\ takes\ the\ form\ of\ an\ interval>$. Additionally, her actions can be interpreted as an act of representing, since she follows the theorem-in-action $<MA/r: Typical\ represents\ the\ always\ changing\ temperatures>$.

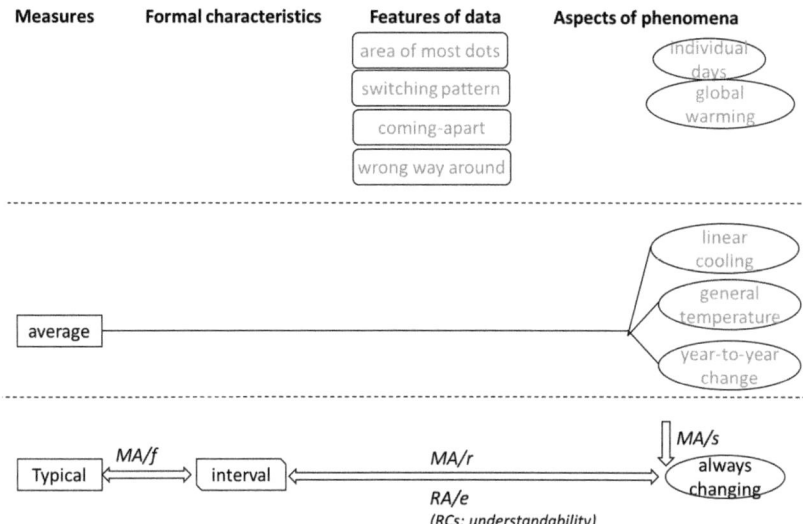

Fig. 7.15: Maria and Natalie start to develop the situative measure Typical by identifying a new aspect of the phenomenon. They also engage in representing and formalizing, as well as in the reflective activity of explicating aims and purposes

Analysis of reflective activities. The students are engaged in the reflective situation of advocating measures. In this scene, Natalie not only uses Typical to represent $||MCs/a:\ always\ changing||$ temperatures, but instead explicitly justifies the choice of Typical (using an interval you can "simply understand it better, because it changes itself from day to day", #738). This explicit reference can be interpreted as an act of explicating aims and purposes of measures ('RA/e'), since the relation between measure and aspect of the phenomenon becomes clear. Underlying this activity is her reflective theorem-in-action $<RA/e:\ Typical$ *as an interval is more easily understandable for representing the always changing temperatures*$>$.

Analysis of reflective concepts. In her reflective theorem-in-action, Natalie draws on the situative reflective concept-in-action of $||RCs:\ understandability||$. She uses this concept to explain that it is not enough that a measure represents an aspect of the phenomenon. Instead, one has to take its purpose in communicating understandings about phenomena into account. Thus, Natalie tackles questions of $||RCg:\ intersubjectivity||$: it is the purpose of measures to provide information that can be understood by everyone in the same way by relying on formal characteristics instead of subjective interpretations of the data. This also fits to her mathematizing activity of formalizing observed in this scene, as in order to be understandable, one has to make efficient use of suitable formal characteristics.

Typical Antarctic Temperatures Problem – Episode 6

A minute later, the students compare different formalizations of Typical on the filled-in report sheets.

III-1a-MN; Phase 5; Start: 46:00
Previously, Maria and Natalie formalized Typical as an interval. Now, they compare Typical to the average. They find that both measures represent different aspects of phenomena.

780	M	Okay. And this one here is, well, the green one. This one *[points to Large Interval Report Sheet]* is sort of indicating. But that's a pretty rough guess. Right?	Achso. Und der hier ist, also der grüne. Der *[zeigt auf Large Interval Report Sheet]* zeigt schon so eigentlich an. Aber das is ja so ziemlich grob geschätzt. Oder?
781	N	Yes. #*(incomprehensible)* Typical. *(incomprehensible)*	Ja. #*(unverständlich)* typisch. *(unverständlich)*
782	M	#Because, then ... then you could this one, if we wanted to help the folks over there. Then, then they could still run around in their normal winter jackets at -8. But at -18 nobody would come along ...	#Weil, dann... dann könnte man ja den, wenn wir den Leuten da helfen sollen. Dann, dann können die bei -8 können die auch noch mit normaler Winterjacke rumrennen. Aber bei -18 will auch keine mehr mit...
783	N	Yes, then you ought to...	Ja da muss man ja...
784	M	That's it, then.	Das ist ja dann.
785	N	Then you'd needed to wear something quite different. That's a little bit too rough then, isn't it? But average temperature isn't really that typical, right?	Dann muss man ja schon was ganz anderes anziehen. Das ist dann auch schon ein bisschen zu grob dann ne? Aber Durchschnittstemperatur ist nicht so wirklich typisch, oder?
786	M	What, what do you mean by typical? Well, of course the average temperature is #what is typical.	Was, wie typisch? Ja natürlich ist die Durchschnittstemperatur #das typische.
787	N	#Well okay, it's what's typical, but? Um.	#Ja okay, ist das typische, aber? Hm.
788	M	Yes.	Ja.
789	N	-15.	-15.
790	M	Okay nah, typical actually means where the mo... most ... nah.	Okay nee, typisch heißt ja eigentlich wo die mei... meist... nee.
791	N	Typical.	Typisch.
792	M	The average temperature actually isn't what's typical. Because that's just in general, overall. #For example, typical for here *[points to 2004]* would be like here *[points to 2004, -14]*.	Die Durchschnittstemperatur ist ja gar nicht das typische. Weil das ist ja einfach nur allgemein, die Ganzen. #Das Typische wär ja jetzt zum Beispiel bei dem *[zeigt auf 2004]* ja hier *[zeigt bei 2004, -14]*.
793	N	#Typical, I believe, it's simply what's the most or most frequent. But I think, like, it can ...	#Typisch glaub ich einfach nur, was so #am meisten oder am öftesten ist. Aber find ich schon, kann...
794	M	#The most. Where the most points, definitely. Frequentest isn't a word. *[laughs]* Most freq...	#Am meisten. Wo die meisten Punkte, auf jeden Fall. Öftesten kann man nicht sagen. *[lacht]* Am oft...

Analysis of mathematizing side. In this excerpt, the students compare the different interpretations of $||MCs/m: Typical||$ on the filled-in report sheets. The discussion again turns towards the similarities and differences of $||MCs/m: Typical||$ and the $||MCg/m: average||$ ("average temperature isn't really that typical", 785 and "of course the average temperature is what is typical", #786). This seems to be an undecided question, until they find a difference in corresponding features of the data. Re-discovering a feature of the data already identified at an earlier point, Natalie finds that <*MA/f: Typical corresponds to the area of most dots*> (#790). Engaging in representing an aspect of the phenomenon also discovered earlier, Maria notes that <*MA/r: The average represents the general temperature*> (#791). Although this mainly posits a difference between features of the data and aspects of the phenomenon, to the students this presents a reason for the difference between the two measures; it seems the students hold that <*MA/s: the area of most dots does not correspond to the general temperature*>.

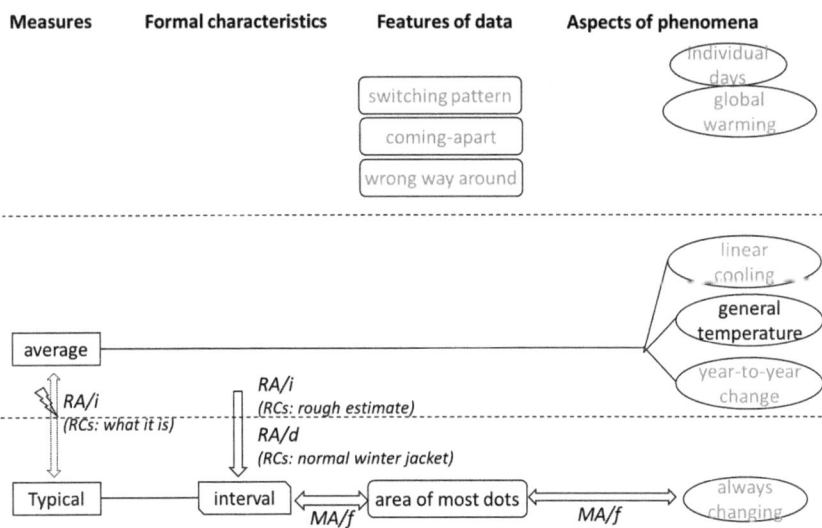

Fig. 7.16: Maria and Natalie find a feature of the data corresponding to Typical. They contrast average and Typical and engage in various reflective activities.

Analysis of reflective activities. In this scene, the students engage in various reflective activities. They first note that Typical is "like a rough guess" (#780). Again, they explain a measure not through any represented aspects, but by providing some other form of meaning. This is an act of identifying patterns of thought with the underlying reflective theorem-in-action <*RA/i: Typical as an interval from -18 to -8 is a rough guess*>.

This presents some limitations for the measure: such a large interval would not be helpful for preparation, as a "normal winter jacket" could not protect from -18°C (#782). Typical should rather reflect that one needs special equipment in Antarctica (you have to wear "something quite different", #785). Thus, the students denominate risks and limits with the underlying theorem-in-action *<RA/d: Typical as an interval from -18 to -8 cannot be used for preparation with a normal winter jacket>* (#782). These considerations about the type of interval to be used for Typical seem to help the students in contrasting the two measures Typical and Average. Contrasting the two measures, they follow the reflective theorem-in-action *<RA/i: the average is not Typical>* (#792).

Analysis of reflective concepts. Describing Typical as an interval from -18 to -8 by drawing on the reflective concepts-in-action of ||*RCs: rough guess*||, the students again find a situative way to address a precursor to the general concept of ||*RCg: perspectivity*|| of measures. If one used this interval, one would get an impression of the phenomenon that would actually be too rough: the temperature range of a ||*RCs: normal winter jacket*|| does not fit to that interval, which could be an individual expression that such an interval has little ||*RCg: contextual relevance*||. This is especially interesting, as the students are able to draw on their experience with the phenomenon of temperatures in order to support their reflective activities. The insights into patterns of thought and risks and limits help the students in contrasting the two measures, as ||*RCs: what Typical is*|| differs from ||*RCs: what the average is*||. In this way, the students make first steps towards the general reflective concept of the ||*RCg: specificity*|| of measures, as the fact that every measure represents only very specific aspects of phenomena and corresponds to specific features of data is discovered by them for these two specific measures. The girls get a first access to the observation that any difference in measures also results in different aspects of the phenomena and features of the data.

Typical Antarctic Temperatures Problem – Episode 7

Some minutes later, the students are challenged by the design experiment leader to explain the difference between average and Typical one last time.

III-1a-MN; Phase 5; Start: 50:50
Previously, Maria and Natalie contrasted the measures of average and Typical. Now, the design experiment leader prompts them to again summarize the differences between average and Typical.

854	DL	And, um, you just said that there's a difference between average and typical.	Und hm ihr habt gerade noch gesagt, da gibt's einen Unterschied zwischen dem Durchschnitt und dem Typisch.
855	M	Okay, yes.	Achso ja.
856	N	Right, because the average is always a single number ...	Ja, weil der Durchschnitt ist ja immer nur eine Zahl...

857	M	If you'd now ...	Wenn man jetzt...
858	N	And the average is also rather inaccurate, because it doesn't tell you anything at all about an individual day. And now about, uh, typical, I'd say that it's, uh like, a range between two numbers, because then you simply can better overlook how it is most of the time.	Und Durchschnitt ist auch so ziemlich ungenau, weil das sagt überhaupt nichts über einen einzelnen Tag aus. Und wenn man jetzt hm Typisch, würd ich eher sagen, dass das so ne hm so ne Spanne zwischen zwei Zahlen ist, weil man dann einfach besser so überblicken kann, wie es meistens so ist.
859	M	Yes, because we said like, you know, 13 or 12 was typical for the first year #and there aren't even any dots near 13 or 12.	Ja, weil wir haben ja auch gesagt irgendwie, ich glaub 13 oder 12 war typisch für das erste Jahr #und bei 13 oder 12 sind überhaupt gar keine Punkte.
860	N	#Yes.	#Ja.
861	M	Um, and then you can't say that there, that's the typical ...	Hm und dann kann man ja auch nicht sagen, dass da, das dort die typischen...
862	N	Yes, that would then be the average then.	Ja das wär dann Durchschnitt dann.
863	M	That would be average and not typical.	Das wär dann der Durchschnitt und nicht das Typische.

Analysis of mathematizing side. In order to explain the differences between average and Typical, the students focus on their formal characteristics. Natalie states that "the average is always a single number, #856), whereas Typical should be a "range between two numbers" (#858). Thus, the students engage in formalizing, following their theorems-in-action $<MA/f$: *the average takes the form of a number>*, instead of an interval (#856). Furthermore, they argue that Typical cannot take the value of 13 or 12, because "there aren't even any dots" (#859). Thus, they identify formal characteristics for Typical and average concerning their concept-in-action of $||MCs/c$: *actual data$||$*. Their formalizing activity is supported by the theorems-in-action $<MA/f$: *Typical corresponds to actual data points>* (it cannot be 13 or 12, because "there aren't even any dots", #859) and $<MA/f$: *the average is not limited to actual data points>* (If there were no dots, it would need to be the average instead of Typical, #862). They finally engage in structuring by identifying the aspect of $||MCs/a$: *How it is most of the time$||$* and arguing that $<MA/r$: *Typical represents how it is most of the time>* (#858).

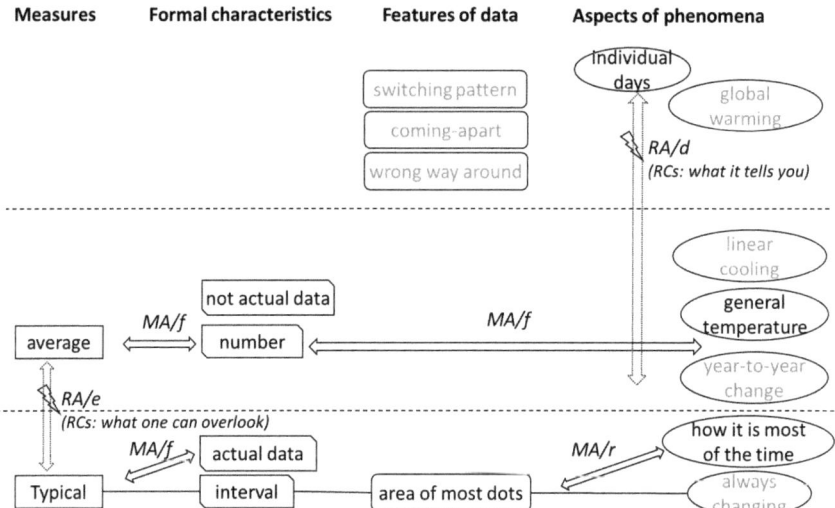

Fig. 7.17: Maria and Natalie heavily engage in formalizing while contrasting average and Typical. They also identify a new aspect of the phenomenon and engage in explicating aims and purposes and denominating risks and limits.

Analysis of reflective activities. Natalie contrasts Typical and average by arguing that the average "doesn't tell you anything at all about an individual day" (#858). Again, this is an act of denominating risks and limits of the average. From her activity, her theorem-in-action of $<RA/d$: *the average does not tell you anything about individual days>* can be inferred. The students also compare the measures by finding that with Typical, one "can better overlook how it is most of the time" (#858). This can be interpreted as an act of explicating aims and purposes, with the underlying theorem-in-action of $<RA/e$: *Typical can be used to better overlook how it is most of the time than the average>*.

Analysis of reflective concepts. In the theorems-in-action, the reflective concept $||RCs$: *what it tells you*$||$ surfaces again in the relationship between $||MCg/m$: *average*$||$ and $||MCs/a$: *individual days*$||$. As before, this might be a way for the students to address the $||RCg$: *perspectivity*$||$ of measures. This also resonates with the students comparing the aims and purposes of the measures by drawing on the situative reflective concept of $||RCs$: *what one can overlook*$||$. The students use this situative concept to address the usefulness of measures: measures are used in statistical investigation, and statistical investigation is conducted in order to provide $||RCg$: *insight*$||$ into phenomena that would not be created otherwise. In this way, the students find a situative precursor that allows them to address that measures expand one's own contextual knowledge, thus creating new insights for the individual.

Effects of design elements for the Typical Antarctic Temperatures Problem

Throughout the design experiment, the students engage in various mathematizing and reflective activities (Fig. 7.18). After initially identifying several features of the data, the students' focus shifts towards aspects of the phenomenon and their relation to the general measure average. They heavily engage in the mathematizing activity of structuring, identifying more and more aspects of the phenomenon. After creating their own report sheet and evaluating the filled-in report sheets, they start to contrast the average with the situative measure Typical. This contrasting is aided by mathematizing activities as well as reflective activities. In the end, they use both measures to represent distinct aspects of the phenomenon and formalize them through different formal characteristics.

Regarding the evaluation of the effects of the design elements, various intended as well as unanticipated effects occurred (Table 7.22). Generally, the design elements worked as intended. The Antarctic weather context allowed the students to engage in the mathematizing activity of structuring, with them identifying aspects of the phenomenon such as ||*MCs/a: individual days*|| or ||*MCs/a: how it is most of the time*|| which show similarities to the intended aspects of ||*MCs/a: short-term variability*|| and ||*MCs/a: typical temperatures*||. Thus, the context of Antarctic temperatures seems to implement DP/Context, as it supported the students mathematizing activities.

The empty and filled-in report sheets prompted the students to engage in formalizing, and rich discussion ensued regarding the formal characteristic of ||*MCs/m: Typical*|| as a ||*MCg/c: number*|| or ||*MCg/c: interval*||. This shows the working of the design principle DP/Formalizations: the students chose an interval as a formal characteristic of Typical in accordance to the aspect of the phenomenon they aimed to represent.

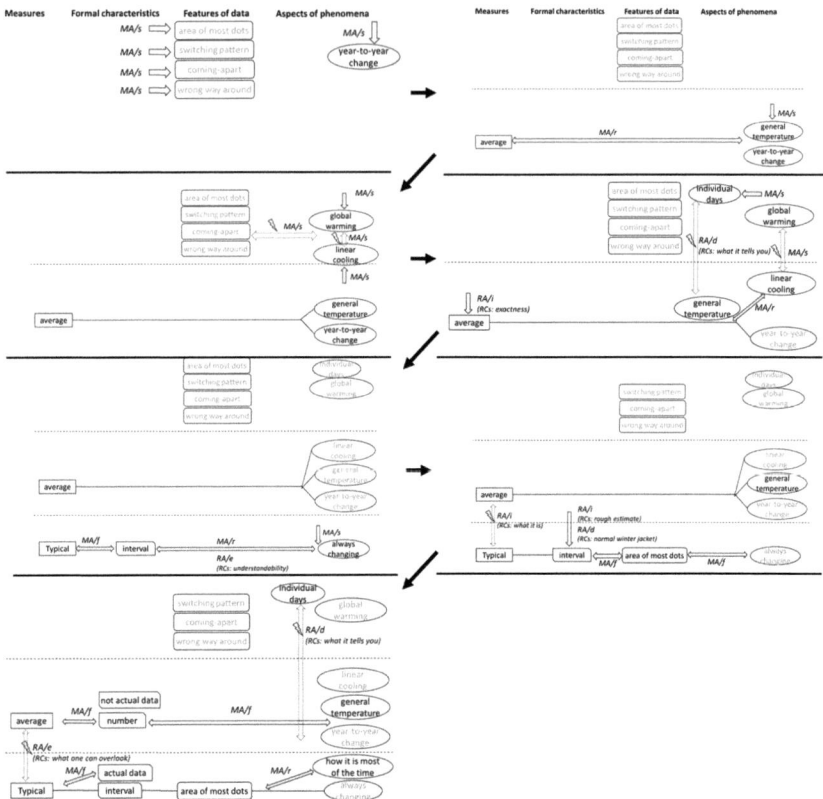

Fig. 7.18: Overview on Maria and Natalie's learning process during the Typical Antarctic Temperatures Problem

The design principle DP/Measures aimed at eliciting mathematizing activities by providing the students with other possible interpretations of measures. The design principle was implemented using the empty and filled-in report sheets and guided the students' learning processes. Aspects of the phenomenon such as ||MCs/a: always changing|| or ||MCs/a: how it is most of the time|| emerged only through the activities of structuring and representing prompted by the examination of the filled-in report sheets. These aspects then played a crucial part for the students' development of Typical.

Tab. 7.23: Intended and actual concepts-in-action identified during the Typical Antarctic Temperatures Problem. Printed bold are concepts-in-action which were not anticipated, but largely supportive to the learning process. Concept-in-action are grouped by category (/m, /c, /f, /a); actual concepts-in-action do not necessarily correspond to intended concepts-in-action

Design element	Intended concepts-in-action	Actual concepts-in-action
Antarctic weather context		*||MCg/m: average||*
	||MCs/a: short-term variability||	*||MCs/a: year-to-year change||*
	||MCs/a: long-term patterns of center and spread||	*||MCs/a: general temperature||*
	||MCs/a: possibility of predicting temperatures||	*||MCs/a: linear cooling||*
		||MCs/a: global warming||
	||MCs/a: typical temperatures||	*||MCs/a: individual days||*
	||MCs/a: exceptional days||	*||RCs: normal winter jacket||*
Single-year data	*||MCs/f: Two modes||*	*(not observed)*
	||MCs/f: Range from -23.5 to -4.5||	
	||MCs/f: Middle 50% from -17 to -9||	
Three-year data	*||MCs/f: Different modes||*	*||MCs/f: coming-apart||*
	||MCs/f: Range from -35 to -4||	*||MCs/f: area of most dots||*
	||MCs/f: Similar Middle 50% for 2002 and 2004||	*||MCs/f: alternating pattern||*
		||MCs/f: wrong way around||
Empty report sheet	*||MCg/m: Minimum||*	*||MCg/m: Minimum||*
	||MCg/m: Maximum||	*||MCg/m: Maximum||*
	||MCs/m: Typical||	*||MCs/m: Typical||*
	||MCg/c: Method of calculation||	
	||MCg/c: Interval||	*||MCg/c: number||*
	||MCs/a: typical temperatures||	
Filled-in report sheets	*||MCg/c: Interval||*	*||MCg/c: interval||*
	||MCg/c: Number||	*||MCg/c: number||*
	||MCs/c: Single highest frequency||	
	||MCs/f: Main body of data||	
	||MCs/f: Area encompassing most modes||	
	||MCs/f: Highest density in 2002 and 2004||	
	||MCg/f: Boundary of data||	
	||MCg/f: Mode of 2003||	
	||MCg/a: Worst observed case||	*||MCs/a: always changing||*
	||MCs/a: Exceptional days||	*||MCs/a: how it is most of the time||*
	||MCs/a: Most likely temperatures||	
	||MCs/a: Reasonable possible temperatures||	
	||MCs/a: Reasonably expected temperatures||	

The design elements also had some interesting unanticipated effects (printed bold in Table 7.22). Although the $||MCg/m: average||$ was never intended to be used by the students, the context of temperatures seems to have prompted the students to draw on their knowledge regarding this general measure. While this provided the students with a familiar way to deal with the data, it also prompted them to identify the unexpected aspect of $||MCs/a: general temperature||$. This seems to have pushed other intended aspects into the background, as $||MCs/a: always changing||$ and $||MCs/a: how it is most of the time||$ only emerged later through the filled-in report sheets.

Instead, the students used the average to identify the $||MCs/a: linear cooling||$ of temperatures. At this point, the context had another unexpected effect: the students drew on their knowledge of $||MCs/a: global warming||$. It is interesting to note that global warming, although implied through the context, does not actually have any representation in the data, at least from a scientific standpoint. The students could only draw on their personal knowledge in order to identify this aspect of the phenomenon. Although their knowledge of global warming was informal and scientifically incorrect, they could productively use it to refute the idea of a $||MCs/a: linear cooling||$ of temperatures.

A third unexpected effect of the context could be observed in the students' reflective activities. The students drew on the situative reflective concept-in-action of $||RCs: normal winter jacket||$. This concept is bound to the context, and only their knowledge of temperatures enabled them to engage in the denomination of risks and limits in this situative way. Thus, DP/Context seems to also possibly support students' reflective activities by providing a situative language students can intuitively use for their reflections.

Arctic Sea Ice Problem – Episode 1

The design experiments concerning the Arctic Sea Ice Problem took place a week later. During the course of the design experiment, Maria and Natalie continue to develop their measure Typical. The main part of their activity takes place during Phase 3. Since not much concerning Typical is added during the other phases of the design experiment, these phases are left out of the analysis.

The analysis starts in the beginning of Phase 3. Previously in Phase 2, Maria and Natalie interpreted the filled-in report sheets for Arctic sea ice from 1982 to 2012, without having access to the data of 2012 themselves (Fig. 7.7 and Fig. 7.9). These filled-in report sheets gave different summaries, two report sheets asserting that almost no change in ice took place, one report sheet that the ice dramatically decreased. Because the students did not have access to the data of 2012, this was intended to prompt them to investigate the different situative measures used. Maria and Natalie, however, noticed these differences, but could not find any reasons for those differences.

In the beginning of Phase 3, the students acknowledge a changing Arctic sea ice, but still do not attribute the different summaries of the filled-in report sheets to the different situative measures employed. This analysis begins as the design experiment leader prompts them to investigate the use of Typical. In this scene, Maria and Natalie question the use of a single number as the formal characteristic of Typical, as used in the Typical Number Report Sheet. As the students already did in the previous design experiment session, they again contrast average and Typical.

III-2a-MN; Phase 3; Start: 23:00
Previously, the students were given the previously missing data of Arctic sea ice from 2012. Now, the design experiment leader prompts them to focus on the use of Typical.

284	DL	Okay. Um, last time we also talked about typical, I believe.	Okay. Ähm, letztes Mal haben wir uns auch noch über so typisch unterhalten, meine ich.
285	M	Um.	Mhm.
286	DL	I believe.	Meine ich.
287	M	What's typical for...	Was typisch für den...
288	DL	And now typical was also partly drawn here *[points to report sheets using Typical]*, did you, do you find this is helpful or not? Well, ...	Und jetzt wurde hier *[deutet auf Steckbriefe mit Typisch]* auch teilweise typisch eingezeichnet, fandet, findet ihr das hilfreich oder nicht? Also...
289	M	Typical, hang on, it says typical is 14, right?	Typisch wart mal, da steht typisch ist 14 oder?
290	N	Yes *#(incomprehensible)*.	Ja. *#(unverständlich)*
291	M	#Typical – uh, but why does it say 13 is typical here *[points to Typical Number Report Sheet]*? From minimum and...	#Typisch – hä aber wieso steht da dann typisch 13 *[deutet auf Typischer-Wert-Steckbrief]*? Von Minimum und...
292	N	Uh, that can't be that there *[points to diagram, 13 million km²]* is typical, because...	Hä kann doch gar nicht sein, dass da *[deutet auf ca. 13 Mio. km²]* typisch 13 ist, weil ...
293	M	Okay...	Ja na...
294	N	Typical is actually like a range, right? Between #this and that, because...	Typisch ist doch eigentlich so ne Spannweite oder? Zwischen #das und das, weil...
295	M	#Well, no – yes typical as in our typical, again is – hang on – typical – um, does typical refer to one, like should you say one degree-number that is like – the average again and that would be typical, or the #range?	#Ja ne – ja typisch für unser typisch, ist ja wieder – warte – typisch – mhh bezieht die sich auf typisch auf ein also soll man dann eine Gradzahl sagen, die so – wieder der Durchschnitt wär und das ist dann typisch wär oder die #Spannweite?
296	N	#Well I'd say range – but...	#Ja ich würd sagen Spannweite – aber...
297	DL	What would, exactly, what would you say should be picked?	Was würde, genau, was würdet ihr sagen sollte man denn nehmen?
298	N	Well I'd certainly pick range, because that simply tells you more.	Also ich würd auf jeden Fall Spannweite nehmen, weil das einfach viel mehr aussagt.
299	M	That also #tells you more.	Das sagt #auch viel mehr aus.

300	N	#Because you couldn't say that it's typically 11 degrees.	#Weil man kann ja nicht sagen, es ist 11 Grad typisch.
301	M	*(incomprehensible)*	*(unverständlich)*
302	N	Yes, you can now - well, okay, if now we'd say degrees, - that really isn't - that would be the average and that would then not be typical, because typical would somehow be - it is between 9 and 12 degrees or like.	Ja man kann es jetzt – ja okay, wenn wir jetzt mal Grad sagen würden, - das ist ja nicht – das wär dann ja der Durchschnitt und das wär ja dann nicht typisch, weil typisch wär dann irgendwie – es ist zwischen 9 und 12 Grad oder so.

Analysis of mathematizing side. Only after the design experiment leader gives the prompt to focus on Typical (#288) do the students seem to notice the different formal characteristics of the situative measure $||MCs/m\!:\ Typical||$ on the filled-in report sheets. The formalization of Typical as a $||MCg/c\!:\ Number||$ on the Typical Number Report Sheet seems to irritate the students (#289). Although in the first design experiment, they invented that $<MA/f\!:\ Typical\ takes\ the\ form\ of\ an\ Interval>$, this seems now to again require justification (it is "actually like a range", #294 but could also be a number, #295). Although the students use the term 'range' they do not use it to refer to the general measure by the same name, but rather appropriate the term as an individual name for 'interval'. When comparing $||MCs/m\!:\ Typical||$ to the $||MCg/m\!:\ Average||$, they again come to the conclusion that $<MA/f\!:\ takes\ the\ form\ of\ an\ interval>$ (#298, Typical should be "between 9 and 12 degrees or like", #302).

It is interesting to note that, although the phenomenon to be investigated is Arctic sea ice, the students use language related to temperatures (#300, #302): they draw on the concept-in-action regarding the aspect of Antarctic temperatures of $||MCs/a\!:\ degrees||$ in order to explain Typical ("it is between 9 and 12 degrees", #302). This could be a first sign of the students broadening the contextual neighborhood of their situative measure $||MCs/m\!:\ Typical||$, as they begin to draw connections between two different contexts.

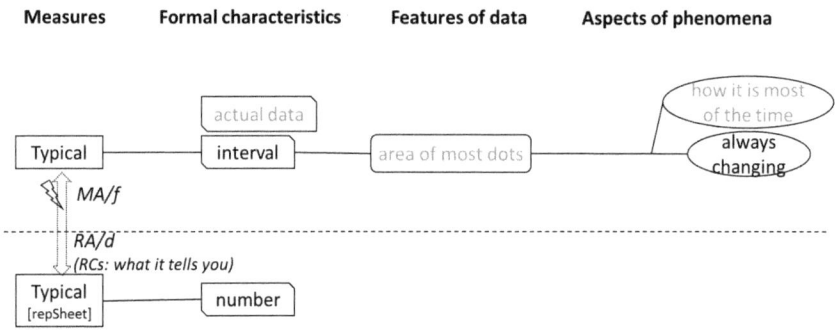

Fig. 7.19: Maria and Natalie contrast their own situative measure Typical with the one of the Typical Number Report Sheet through formalizing and denominating risks and limits

Analysis of reflective activities. In order to choose between two possible formal characteristics for Typical, namely of $||MCg/c: number||$ or $||MCg/c: interval||$, Natalie finds that choosing a number would limit the use of Typical (an interval "tells you more", #298). Beyond this utterance, the following theorem-in-action can be identified: $<RA/d:$ *Typical as a number does not tell you enough>*, which is an indicator of her reflective activity of denominating risks and limits. This reflective activity seems to support her mathematizing activity of formalizing Typical, so she prefers choosing an interval as a formal characteristic for Typical (#298).

Analysis of reflective concepts. In her reflective theorem-in-action, Natalie draws on her situative reflective concept-in-action $||RCs:$ *what it tells you*$||$. This is a situative reflective concepts already used during the Typical Antarctic Temperatures Problem. It thus seems that situative reflective concepts can have some stable qualities and do not only appear disconnected from previous situations. Again, this could be a situative way to address the $||RCg:$ *perspectivity*$||$ of measures, which in this case can also aid the mathematizing activity of formalizing.

Arctic Sea Ice Problem – Episode 2

After the students take some time to try to visually identify Typical in the data, the design experiment leader intervenes and asks them for the usefulness of Typical. The students again contrast Typical with the average.

III-2a-MN; Phase 3; Start: 27:00			
Previously, the pair settled on the formalization of Typical through an interval. Now, the design experiment leader asks them whether they would actually use Typical on a report sheet. The students contrast average and Typical.			
357	DL	And would you get anything out of this? Like, assuming someone wrote this down in one of these sheets *[points to report sheets]*.	Und würde euch das jetzt irgendwas bringen? Also wenn angenommen jetzt schreibt das jemand so auf so einen Steckbrief *[deutet auf die Steckbriefe]*.
358	N	Well?	Ja?
359	DL	Would this, would this be sufficient, or ...	Würd das, würd das reichen oder ...
360	N	No.	Nein.
361	M	No.	Nein.
362	M	#Well, somehow yes.	#Ja irgendwie schon .
363	N	#It also wouldn't be sufficient. I think average is still important, right?	#Also reichen würde das nicht, ich finde Durchschnitt ist schon wichtig, oder?
364	M	Um, yes.	Mhh ja.
365	N	Just range, like, what is typical, it's ...	Bloß Spannweite, also was so typisch ist, ist ja...
366	M	# Isn't necessarily ...	#Ist ja nicht unbedingt ...

367	N	#It simply tells you , um, more about the individual, ehm, day than if you say the average, because if you...	#Das sagt einfach auch,ähm, so mehr über die einzelnen, ähm,Tage aus als wenn man den Durchschnitt sagt, weil wenn du ...
368	M	#Average tells you...	#Durchschnitt sagt ja...
369	N	#Have an average of like 12, then on a given day it also can be #18 degrees or	#Durchschnitt irgendwie zu 12 hast, dann kann es an einem Tag aber auch #18 Grad oder
370	M	#Minus 17 – yes *[incomprehensible]*.	#Minus 17 – ja *[unverständlich]*.
371	N	Or minus 10 or so degrees.	Oder Minus 10 oder so – Grad sein.
372	DL	Mhm.	Mhm.
373	N	And the average just tells you what happened in general, but I think the range just tells #you better what	Und der Durchschnitt sagt dann einfach so aus, was insgesamt passiert ist, aber ich glaub die Spannweite sagt #einfach viel besser aus, was so
374	M	#typical.	#typisch
375	N	Was in general.	Insgesamt war

Analysis of mathematizing side. The students again contrast $||MCs/m: Typi-cal||$ with the $||MCg/m: Average||$. Again, this is a question of the formal characteristics of $||MCg/c: number||$ or $||MCg/c: interval||$ (which the students call "range", #365, appropriating the language offered through the scaffolding of the filled-in report sheets). In order to characterize the differences, Natalie first engages in structuring by reconnecting the aspect of $||MCs/a: individual days||$ (#367), which first was identified for Antarctic temperatures. She also engages in representing through her theorem-in-action that $<MA/r: the average does not represent always changing temperatures>$ (the average does not "tell" about deviations from the average, #369, #371). Through this activity of representing, she can describe the differences between the measures: Natalie distinguishes between what "happened in general" (#373) and what "was in general" (#373). Although on a surface level her words do not seem different, to her, this seems to present a significant difference. Based on the students' previous uses of the average, her theorems-in-action are reconstructed as $<MA/r: the average represents the general temperature>$ and $<MA/r: Typical represents individual days and the always changing temperatures>$ - the latter of which can be interpreted as the pattern of the spread instead of the center (the deviations from the average in #367, #369).

Measures   Formal characteristics   Features of data   Aspects of phenomena

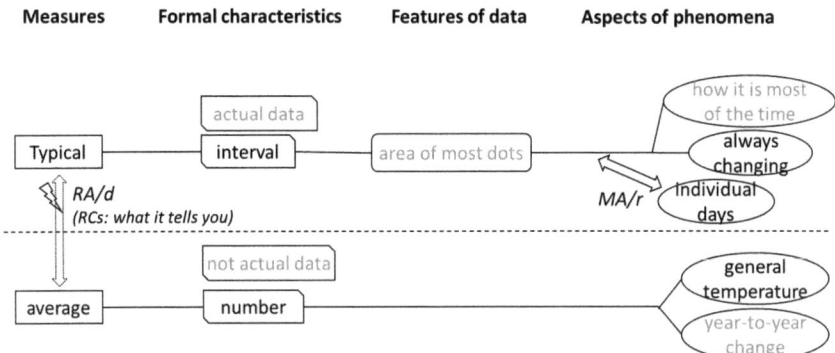

Fig. 7.20: Maria and Natalie contrast the two measures by denominating risks and limits and engaging in representing

Analysis of reflective activities. The comparison between the measures again is articulated by contrasting the different aspects of the phenomenon addressed of the two measures. This can be interpreted as an act of denominating the limits of the average with the underlying theorem-in-action *<RA/d: The average does not tell you anything about individual days and the always changing temperatures>*, since these aspects of the phenomenon, however, would be important to represent for creating report sheets (#367).

Analysis of reflective concepts. The students again draw on their concept-in-action of ||*RCs: What it tells you*|| to frame the differences of the measures. It is worth noting that the students do not give any preference to one measure over the other – each measure "simply" represents other aspects of the phenomenon (#373). Their situative reflective concept of ||*RCs: What it tells you*|| thus is a likely precursor to the general reflective concept of ||*RCg: perspectivity*||.

Arctic Sea Ice Problem – Episode 3

After having criticized the formalization on the Typical Interval Report Sheet for being unintelligible, Maria and Natalie find a procedure of calculation for their own situative measure Typical in the following scene.

III-2a-MN; Phase 3; Start: 29:00
Previously, the students noticed that the two intervals given in the Typical Interval Report Sheet do not seem to follow a consistent method of calculation, since the values for 1982 and 2012 are nearly the same although the minimum dropped considerably. Now, The design experiment leader picks up that criticism but challenges the students to justify their own formalization of Typical. Drawing on their knowledge of the average, they develop a procedure of calculation for Typical.

| 400 | DL | But you also did it one way and then another, didn't you? With your typical. | Aber ihr habt jetzt auch mal so mal so gemacht oder? Bei euren typischen. |

401	M	Yes.	Ja.
402	N	Well I don't know either how to calculate the typical. #I believe...	Ich weiß auch nicht wie man die typischen ausrechnet. #Ich glaub...
403	M	#I don't know either how to do this.	#Ich weiß auch nicht wie man das macht.
404	N	You just say, um, um, well you, I believe you start from the average and then you look at what is the highest and what is the lowest temperature. And from that you then determine the middle value, like between, I'd say, between #average and...	Man sagt einfach so, ähm,–, ähm,– also man – ich glaub man geht vom Durchschnitt aus und dann guckt man, was die höchste und was die niedrigste Temperatur ist und davon muss man dann so einen Mittelwert finden, oder, also zwischen ich würde mal sagen zwischen #Durchschnitt und...
405	M	#average and...	#Durchschnitt und...
406	N	And the #lowest temperature.	Und der #kleinsten Temperatur.
407	M	#lowest and the hightest – yes and the highest or...	#kleinsten und der größten – ja und der höchsten oder...
408	N	#Yes, well first...	#ja also erstmal...
409	M	#I think we've been talking all this time about temperatures *[points to diagram]*, but these aren't even #temperatures.	#Ich find wir reden die ganze Zeit über Temperaturen *[deutet auf Diagramm]*, aber das sind keine #Temperaturen.
410	N	#Yes I know, but if we #just said temperatures...	#Ja ich weiß, aber wenn wir #jetzt halt Temperaturen sagen...
411	DL	#I know what you mean.	#Ich weiß, was ihr meint.
412	N	Yes, then you take the average and the coldest and from that you again take #the average.	Ja, dann nimmst du den Durchschnitt und die kälteste und machst davon nochmal #den Durchschnitt.
413	M	#The average...	#den Durchschnitt...
414	N	#And then...	#Und dann...
415	M	#And that...	#Und das...
416	N	#From the average and the highest temperatures the average...	#Von dem Durchschnitt und der höchsten Temperatur den Durchschnitt...
417	M	#From – average – highest – average – yes I know...	#Von – Durchschnitt – höchsten – Durchschnitt – ja ich weiß...
418	N	And then the average of the other average is the typical.	Und dann ist der Durchschnitt vom anderen Durchschnitt die typische.

Analysis of mathematizing side. Previously, the students noticed a special characteristic of the Typical Interval Report Sheet: the calculation of Typical for each year on the report sheet does not follow a consistent method. This detail is picked up by the students, showing their careful and thorough involvement with the task design. Initially, the challenge by the design experiment leader to provide a better method seems to be a cause of concern to the students ("Well I don't know either how to calculate the typical", #402). Natalie, however, quickly finds a formalization for $||MCs/m: Typical||$, which both students then develop together.

As determined earlier, a formal characteristic of their situative measure $||MCs/m: Typical||$ is an $||MCg/c: Interval||$, the boundaries of which now need to be determined. Natalie proposes that one starts by finding the $||MCg/m:$ *average*$||$, which then partitions the data into two halves ("you start from the average, and then you look at what is the highest and what is the lowest temperature", #404). From both of these halves, one needs to find a "middle value" (#404). This middle value seems to correspond to the aspect of $||MCs/a:$ *general temperature*$||$ represented by the $||MCg/m:$ *average*$||$. Natalie thus proposes to calculate the average between $||MCg/m:$ *average*$||$ and $||MCg/m:$ *minimum*$||$ (#404, #406), and to do likewise with the $||MCg/m:$ *maximum*$||$. The two new averages then are the boundaries of $||MCs/m:$ *Typical*$||$ (#418). In this way, they are able formalize Typical by giving an explicit method of calculation for $||MCs/m:$ *Typical*$||$.

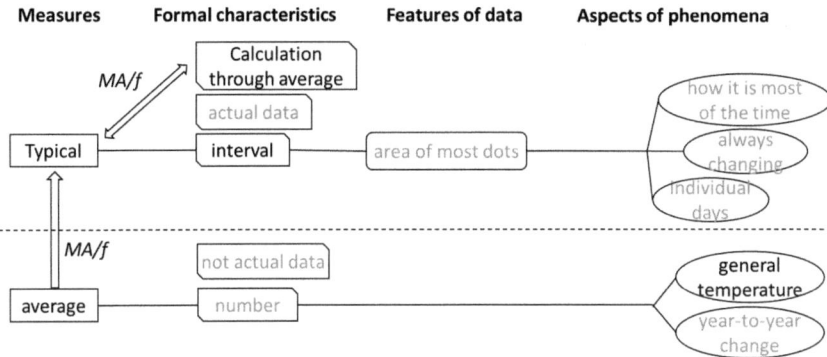

Fig. 7.21: Maria and Natalie engage in formalizing, using the average to find a procedure of calculation for Typical

Effects of design elements for the Arctic Sea Ice Problem

Throughout the design experiment session, the students' conceptual network does not show as many changes as observed in the first design experiment session (Fig. 7.22). The students begin by contrasting their own use of Typical with that on the Typical Number Report Sheet. Later, they begin to contrast Typical with the average, as they already did during the first design experiment session. This helps them in developing their measure Typical, as they formalize it by finding a procedure of calculation.

Measures   Formal characteristics   Features of data   Aspects of phenomena

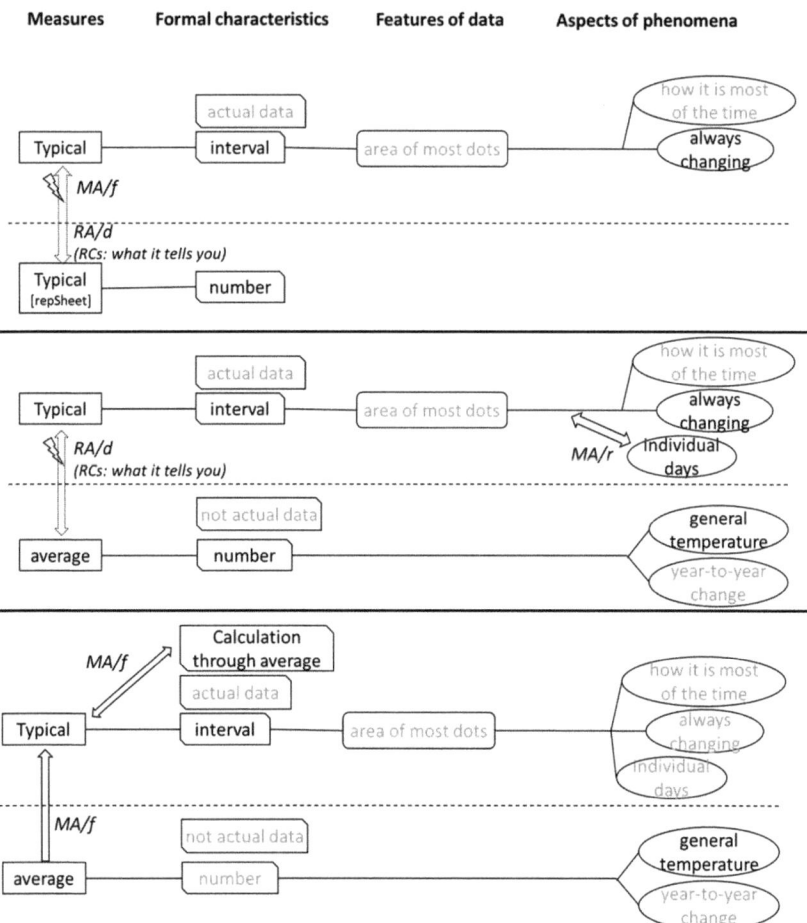

Fig. 7.22: Overview of Maria and Natalie's learning process during the Arctic Sea Ice Problem

Again, different effects of the design elements could be observed during Maria and Natalie's learning process (Table 7.23). Regarding the design principle DP/Formalizations, the design elements mostly succeeded. The design principle intended to elicit activities of formalizing by providing different formalizations of situative measures; it could be observed that the filled-in report sheets prompted the students to identify an interval as formal characteristic for Typical, and even gave an unexpectedly thorough method of calculation utilizing the already familiar general measure average.

Tab. 7.24: Intended and actual concepts-in-action stimulated through the design elements of the Arctic Sea Ice Problem. Printed in bold are unanticipated concepts-in-action.

Design element	Intended concepts-in-action	Actual concepts-in-action
Arctic sea ice con- text	||*MCs/a: winter months*|| ||*MCs/a: summer months*|| ||*MCs/a: melting process*||	||*MCs/m: average*|| ||***MCs/a: degrees***|| ||***MCs/a: individual days***|| ||***MCs/a: changing temperatures***||
Two-year data	||*MCs/f: modes*|| ||*MCs/f: left-skewed*|| ||*MCs/f: spread from 7.5 to 16.5*||	
Three- year data	||*MCs/f: increased spread in data*|| ||*MCs/a: increasing severity of ice melt*||	*(no attention given to features of the data)*
Empty report sheet	||*MCg/m: range*||	||***MCs/m: Typical***|| ||***MCs/c: Calculation through average***||
Filled-in report sheets	||*MCg/m: range*|| ||*MCs/m: Typical*|| ||*MCg/c: number*|| ||*MCg/c: interval*|| ||*MCg/c: calculation through min. and max.*|| ||*MCs/f: main body of data*|| ||*MCg/f: spread of data*|| ||*MCs/a: beginning of winter*|| ||*MCs/a: severity of ice melt*||	||*MCs/m: Typical*|| ||*MCg/c: number*|| ||*MCg/c: interval*||

An ambiguous picture is painted by the design elements concerning DP/Context, which aimed at supporting the students' structuring activities. The context of Arctic sea ice prompted the students to make recourse on aspects of Antarctic temperatures. Although making such explicit connections between the contexts was not anticipated at this point, this allowed the students to overcome the unfamiliar context, to engage with the problem, and to formalize their situative measure Typical. It, however, also resulted in ill-fitting activity of structuring of the phenomenon, as they did not actually structure the phenomenon of Arctic sea ice. They also mostly ignored the features of the data, resulting in a formalization of Typical that relies on formal characteristics and ill-fitting represented aspects of phenomena, but not on corresponding features of the data. In this way, the students' resulting development of Typical does not strictly adhere to the intended hypothetical learning trajectory (cf. Tab. 7.24). Nevertheless, in the end, the students show a sophisticated use of measures, being able to consciously choose between average and Typical according to the phenomenon under investigation.

The focus on Antarctic temperatures also has effects on the influence of DP/Measures. Since the students did not identify genuine aspects of Arctic sea ice, the value of using the general measure range to represent the aspect of the melting process of sea ice was lost on the students, as the students did not identify the represented aspect of the melting process. Thus, they mostly ignored the idea of using the range as a measure.

7.2.2 Overarching effects for Maria and Natalie's learning processes

Throughout the design experiments, Maria and Natalie engage in various mathematizing and reflective activities while developing Typical. Some phenomena observed during the design experiments merit closer inspection.

Identifying aspects not found in data

During Phase 3 of the Typical Antarctic Temperatures Problem, Maria and Natalie identify the aspect of ||*MCs/a: Global warming*||. This structuring activity is not initiated by the data at hand, since the data do not show any features corresponding to an increase of temperatures. Instead, Maria and Natalie draw on their contextual knowledge to identify the aspect. This would otherwise be a helpful strategy: the aspect of ||*MCs/a: how it is most of the time*|| also has its roots in the pairs' contextual knowledge. Whereas the aspect ||*MCs/a: how it is most of the time*||, however, can easily be connected to the feature ||*MCs/f: area of most dots*||, this cannot be done with the aspect of ||*MCs/a: Global warming*||, prompting difficulties with the activities of structuring and representing. In this way, contexts implementing DP/Context can introduce unanticipated possible obstacles for the learning processes. These obstacles, however, could also be used productively: Maria and Natalie used their contextual knowledge to validate their own use of measures. Additionally, further instruction could capitalize on their intuitive knowledge, as it would be worthwhile for them to learn that global warming is a process too subtle to be observed by the naked eye during a window of 13 years.

Volatility of situative concepts without corresponding measures

Also in Phase 3 of the Typical Antarctic Temperatures Problem, Maria and Natalie identify features of the data such as the ||*MCs/f: Area of most dots*|| and ||*MCs/f: Coming-apart*|| that could have provided starting points for developing general concepts such as ||*MCg/f: Spread*||. However, they did not link these features of the data to any corresponding measures, so that when the pair starts using the familiar ||*MCg/m: Average*||, they lose track of those features (Fig.

7.18). Only later, when $||MCs/m:$ *Typical*$||$ is introduced by the report sheets, do they re-discover the feature of the $||MCs/f:$ *area of most dots*$||$, which then leads to further mathematizing activities. It seems that already identified features of the data and aspects of the phenomenon without corresponding measures are volatile, and run the risk of getting pushed into the background by already consolidated general measures. Thus, using general measures such as the average can also suppress the fleeting insights into aspects of phenomena and features of the data. Fitting situative measures, however, can be used to counter the volatility of situative concepts.

Developing by contrasting measures

During most phases, Maria and Natalie use the general measure $||MCg/m:$ *Average*$||$ for structuring the phenomenon through familiar aspects such as $||MCs/a:$ *general temperature*$||$. By contrasting formal characteristics and represented aspects of phenomena of $||MCg/m:$ *average*$||$ and $||MCs/m:$ *Typical*$||$, they are able to develop their own situative measure Typical. This happened even during the Arctic Sea Ice Problem, where DP/Measures implemented by the filled-in report sheet was designed to prompt comparison between a variety of different situative measures. As such, this comparison with the average was not expected by the design, yet proved extremely important for Maria and Natalie's learning processes.

Coincidence of mathematizing and reflective activities

As Figure 7.23 shows, mathematizing and reflective activities coincided during large parts of both design experiments. In part, this is because denominating risks and limits of $||MCg/m:$ *Average*$||$ or $||MCg/m:$ *Typical*$||$ was one way to contrast the two measures. As outlined above, the contrasting of the measures supported the development of Typical. This result presents an interesting opportunity: the findings suggest that developing mathematizing and reflective concepts might not be two separate goals to be pursued independently from each other. Instead, it appears that reflective activities can support the mathematizing activities and thus be helpful, or even an integral, to developing measures. But not only can reflective activities support mathematizing activities, the reverse might also be true. Whereas reflection in literature often is treated as an addendum that should be considered after mathematics has been carried out, the findings show that reflection can take place in the middle of mathematizing activities. Reflection and mathematizing thus show an intertwined nature and can go hand in hand for the development of measures.

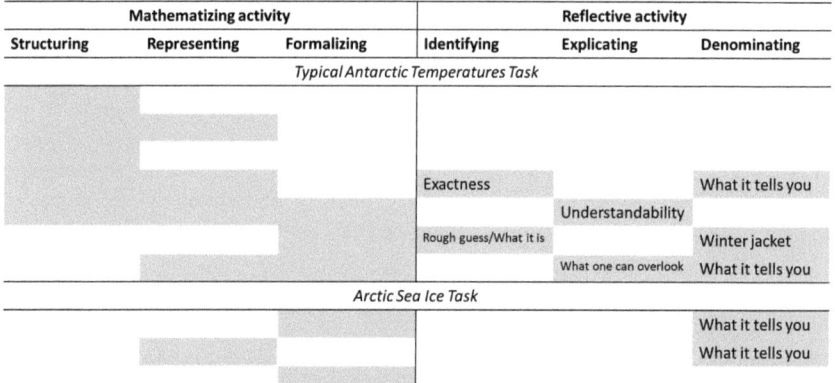

Mathematizing activity			Reflective activity		
Structuring	Representing	Formalizing	Identifying	Explicating	Denominating
		Typical Antarctic Temperatures Task			
			Exactness		What it tells you
			Understandability		
			Rough guess/What it is		Winter jacket
			What one can overlook		What it tells you
		Arctic Sea Ice Task			
					What it tells you
					What it tells you

Fig. 7.23: Maria and Natalie's mathematizing and reflective activities. Highlighted in grey are the pursued activities in each scene. For the reflective side, the reflective concepts-in-action are given

The role of context for the development of Typical

Fig. 7.23 also shows the importance of the design principle DP/Context. During large parts of the design experiments, Maria and Natalie engaged in the mathematizing activity of structuring. This activity was directly supported by the contexts of Antarctic Temperatures and Arctic sea ice. However, context did not only support the activity of structuring, but also in some cases the other mathematizing activities. Structuring often occurred along with other mathematizing activities. Additionally, during the Arctic Sea Ice Problem, the use of well-known language surrounding temperatures helped Maria and Natalie to talk about the less familiar Arctic sea ice – although this also caused additional problems, as the students did not identify aspects exclusive to the phenomenon of Arctic sea ice.

The role of context for reflective activities

The context also influenced Maria and Natalie's reflective activities. Some reflective activities coincided with the activity of structuring (Fig. 7.23). Other times, the reflective concepts were directly influenced by the context. In drawing on the concept-in-action ||*RCs: normal winter jacket*|| to denominate risks and limits, the pair heavily draws on their knowledge of temperature phenomena. Apart from this direct influence, because of the importance of the activity of structuring for the development of measures, the coincidence of reflective and mathematizing activities also implies an important function of the context to students' reflective activities.

Richness and stability of reflective concepts

Fig. 7.23 also shows the different reflective concepts. On the one hand, this reveals a richness in reflective concepts that Maria and Natalie use in their various reflective activities. On the other hand, some reflective concepts appear to be stable: denominating risks and limits is often pursued through the concept of ||*RCs: What it tells you*||, a situative precursor to ||*RCg: perspectivity*||. Not only does this concept often appear in action, but it also appears throughout both design experiments. Such situative concepts thus do not seem to be random, but rather to be part of Maria and Natalie's network of conceptual knowledge, raising hopes for the possibility of systematic development of students' reflective concepts.

7.2.3 Quanna and Rebecca: Developing with a focus on formalizing

The second part of the analysis follows the learning processes of the students Quanna and Rebecca. Compared to Maria and Natalie, the students start out a bit shy and take some time to familiarize themselves with the situation. Their learning processes are characterized by a lower level of interaction and the occasional silence. They seem to lack for words for describing data, often causing them to trail off mid-sentence. Nevertheless, they share ideas and engage with each other, and by the time of the second design experiment, they show increased enthusiasm.

Typical Antarctic Temperatures Problem – Episode 1

The analysis starts in the middle of Phase 2. Previously, the students have drawn the dots of their prediction, whispering to each other to coordinate their proceedings.

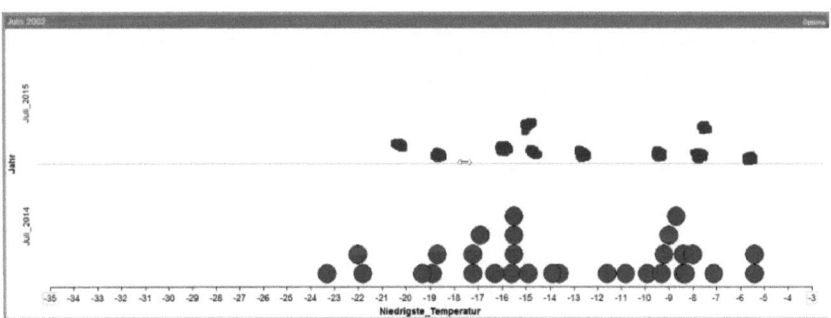

Fig. 7.24   Quanna and Rebecca's prediction from Phase 2

III-1b-QR; Phase 2; Start: 14:50
Previously, Quanna and Rebecca quietly created their prediction for 2015. Now, the students are prompted to explain.

94	DL	Okay. Why do you place the dots where there were the most?	Okay. Wieso tut ihr denn da Punkte hin, wo am meisten waren?
95	R	*[shrugs]*	*[Zuckt mit den Schultern]*
96	Q	Because...	Weil...
97	DL	Or could they be placed elsewhere, or is that not important? Or was this just a guess?	Oder kann man die auch woanders hinsetzen, oder ist das nicht so wichtig? Oder war das nur so geraten?
98	R	A guess, actually. Yes.	Geraten eigentlich. Ja.
99	Q	Actually, no.	Eigentlich ja nicht
100	R	Yes we did. Well, a bit. A bit.	Ja doch. Also geht. Geht so.
101	Q	Because like the risk is relatively high, if there was like ...	Weil also halt das Risiko ist ja relativ hoch, wenn da wo halt...
102	R	Last time the temperatures had occurred so often, that they're again this year, because - because where there the temperature was only once, is like the probability is lower, that it has that temperature again.	Letztes Mal schon die Temperaturen so häufig vorkommen, dass die dieses Jahr auch wieder sind, weil – weil da wo jetzt nur einmal die Temperatur war, ist halt die Wahrscheinlichkeit weniger, dass die da wieder die Temperatur hat.

Analysis of mathematizing side. In this scene, Quanna and Rebecca explain their prediction of 2015 (Fig. 7.24). Previously, Quanna and Rebecca already engaged in structuring the phenomenon by identifying the feature of the distribution of the $||MCs/f:$ *area of most dots*$||$ (#71, not shown). Until now, this feature, however, was left without a corresponding aspect of the phenomenon. When the design experiment leader challenges them to give reasons for their prediction, both students initially are hesitant and take some time to explain their prediction (#98. #99, #100). Finally, by identifying the aspect of $||MCs/a:$ *likely recurring temperatures*$||$ (the "risk" "that it has that temperature again", #101 #102), the students structure the phenomenon by finding a corresponding aspect of the phenomenon to the feature of the $||MCs/f:$ *area of most dots*$||$. Both situative mathematizing concepts, however, are not yet related to any representing measure.

Fig. 7.25: Quanna and Rebecca engage in structuring by identifying a feature of the data and an aspect of the phenomenon, and linking these concepts to each other

Typical Antarctic Temperatures Problem – Episode 2

The design experiment progresses to Phase 3. Immediately as they are given the additional data, the students begin to silently change their prediction without explicating their thoughts. When they are finished, the design experiment leader asks them to explain their changes.

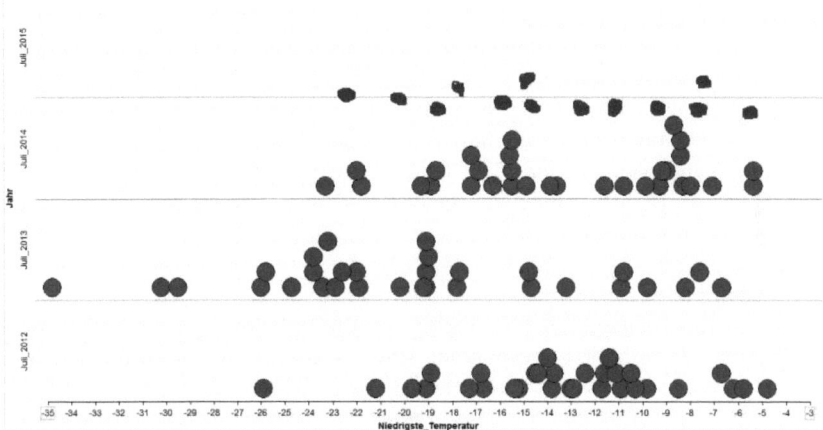

Fig. 7.26: Quanna and Rebecca's prediction for Phase 3

III-1b-QR; Phase 3; Start: 18:00
Previously, Quanna and Rebecca modified their prediction to include the data from 2003 and 2004. Now, they explain their prediction and why they ignored the lower half of the data from 2013.

147	DL	Um. So, why now like this? On the one hand you say you'd ignore those others.	Mhm. So, wieso jetzt so? Also einmal sagst du, die anderen da würdest du nicht so beachten.
148	R	Yes, because here we have nothing really *[points at an area in 2012 from -25 to -35]* and here is still nothing *[points at the same area in 2014]*, so it appeared once only. #And therefore...	Ja, weil da war hier ja noch nichts *[deutet einen Bereich an bei ca. -35° bis -25° Grad im Diagramm von 2012]* wirklich und hier ist ja immer noch nichts *[deutet auf denselben Bereich im Diagramm von 2014]* das ist ja nur einmal gewesen. Und #deswegen...
149	DL	#Um.	Mhm.
150	R	This here all happened more often *[points at middle area across all years]*] so it's more realistic that this - happens again.	Das ist hier ja alles öfters gewesen *[deutet auf den mittleren Bereich in allen Diagrammen]* deswegen ist dann realistischer, dass das – wieder vorkommt.
151	DL	Understood. And these changes - what - what - why did you now change something?	Alles klar. Und diese Änderungen – warum – was – warum habt ihr da jetzt irgendwas verändert?

152	Q	Because, when you count the dots in a column then [indicates vertical lines across the whole diagram] they are like...	Weil wenn man halt die Punkte in einer Spalte zählt dann [deutet vertikale Linien im ganzen Diagramm an] sind die halt ...
153	R	Where like the most [incomprehensible].	Wo dann halt die meisten [unverständlich].
154	Q	Yes.	Ja.

Analysis of mathematizing side. Challenged to explain their prediction (Fig. 7.26), Rebecca starts by explaining that some data entries could be disregarded (dots that "appeared once only", #148). This is reconstructed as Rebecca engaging in structuring by identifying the feature of the data of $||MCs/f$: single occurrences$||$. This feature of the data, however, is not left without interpretation, as it is related to the phenomenon. The feature corresponds to days that will not "happen again" (#150). Thus, Rebecca also identifies the aspect of $||MCs/a$: Unlikely recurring temperatures$||$, engaging in structuring through her implicit theorem-in-action $<MA/s$: Unlikely recurring temperatures correspond to single occurences$||$.

Quanna then engages in formalizing, as she determines a method to find these features: to find the corresponding features, one has to "count the dots in a column" (#152). Thus, she follows her implicit theorem-in-action $<MA/f$: To find single occurrences and the area of most dots, divide the data into columns and count the number of dots in each column$>$. She, however, does not specify how to determine those columns.

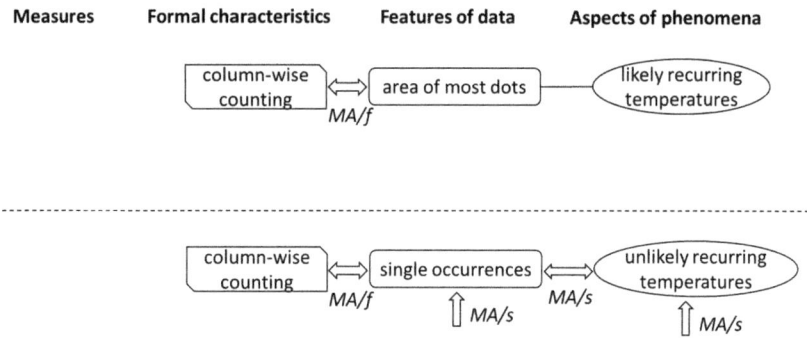

Fig. 7.27: Quanna and Rebecca identify another set of features of the data and aspect of the phenomenon. They engage in formalizing by identifying a procedure of finding the features

Typical Antarctic Temperatures Problem – Episode 3

The design experiment shortly progresses to Phase 4, and the students create their own report sheet, again only whispering to each other (Fig. 7.28). When they are finished, the design experiment leader asks them to explain their own report sheet.

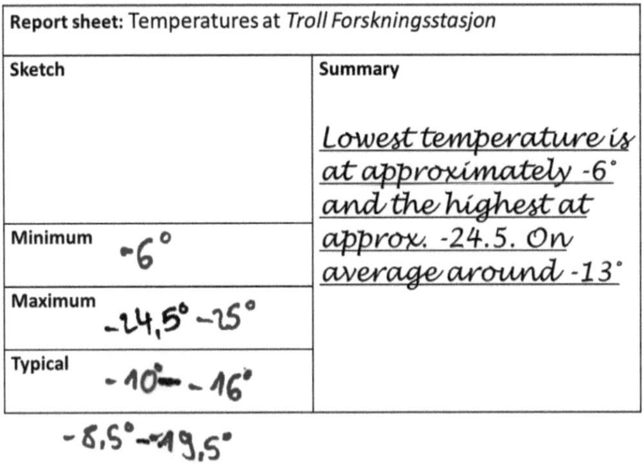

Report sheet: Temperatures at *Troll Forskningsstasjon*	
Sketch	**Summary**
Minimum $-6°$	*Lowest temperature is at approximately -6° and the highest at approx. -24.5. On average around -13°*
Maximum $-24,5° -25°$	
Typical $-10° - -16°$	

$-8,5° -19,5°$

Fig. 7.28: Quanna and Rebecca's own report sheet for the Typical Antarctic Temperatures Task

III-1b-QR; Phase 4; Start: 24:50
Previously, the students created their own report sheet mostly in silence. Now, Rebecca explains the values of the measures on their own report sheet.

220	DL	Okay, awesome. Could you now explain to me again why you wrote all those numbers, -6 degrees and so on? How did you get there?	Okay super. Könnt ihr mir das jetzt nochmal erklären, warum ihr diese ganzen Nummern da aufgeschrieben habt, -6° und so weiter. Wie kommt ihr darauf?
221	R	Because our first dot is roughly at -6 *[points to own prediction, -6°C]*. Then there the last one at -24.5 *[points to own prediction, -24.5°C]*.	Weil unser erster Punkt hier *[deutet auf eigene Vorhersage, -6° Grad]* ungefähr bei -6° liegt. Dann da der letzte bei - 24,5° *[deutet auf eigene Vorhersage, - 24,5° Grad]*.
222	DL	Ah, okay.	Achso okay.
223	R	And - here were most of them *[points to an area from -16 to - 10°C]*, like between 10 and 16 – there were ten here *[points to around -10°C]*, here there are like three at once *[points to same area in 2014]*, and here *[points to area between -18°C and -14°C in 2014]*, and because of that 10 to 16.	Und – hier sind halt die meisten *[deutet auf Bereich von -10 bis -17° Grad]* gewesen, so zwischen 10° und 16° - da waren auch hier 10 *[deutet ca. -10° Grad]* da sind hier so drei auf einmal *[deutet auf dieselbe Stelle von 2014]* und hier *[deutet auf den Bereich zwischen -18° und -14° Grad im selben Diagramm]* deswegen 10 bis 16.
224	DL	Mhm.	Mhm.
225	R	Because here are most *[again points to area from -10°C to -16°C]*, that's why it's typical.	Weil hier auch die meisten sind *[deutet den Bereich von -10° bis -16° Grad nochmal an]*, deswegen ist das typisch.

Analysis of mathematizing side. After briefly explaining their minimum and maximum values (#221, not focus of this analysis), Rebecca turns toward the measure $||MCs/m: Typical||$ on their own report sheet. In order to explain, she engages in the mathematizing activity of representing: $<MA/r: Typical$ *represents the area of most dots>* ("here are the most, that's why it's typical", #225). Thus, the students are able to use Typical to represent a feature of the data they already identified previously in the design experiment, but was until now left without a representing measure.

Looking at the mathematizing activities reconstructed in the design experiment session so far, it already is noticeable that the students mostly engage in formalizing and representing, and only seldom in structuring. Although they already identified the aspect of $||MCs/a: Likely recurring temperatures||$, Rebecca mostly focuses on features of the data for explaining their use of Typical (how many dots there are at once and where the most dots are, #223 #225). This is an unanticipated effect, as it was hoped that the phenomenon of Antarctic weather would inform the choice of measures.

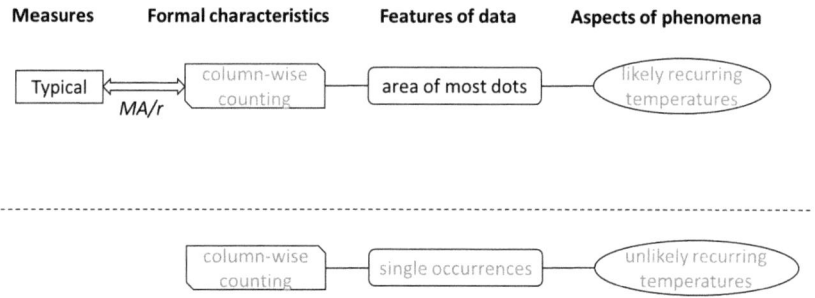

Fig. 7.29: Quanna and Rebecca use the situative measure Typical to represent a feature of the data identified earlier in the design experiment session

Typical Antarctic Temperatures Problem – Episode 4

The following excerpt stems from Phase 5 and starts some minutes after the students took a first look at the filled-in report sheets (Fig. 7.8). Rebecca starts to evaluate the filled-in report sheets, after the design experiment leader had prompted the students to.

III-1b-QR; Phase 5; Start: 31:00
Previously, the students explained their own report sheet, and the design experiment leader gave them the filled-.in report sheets. Now, Rebecca evaluates the Typical Number Report Sheet and identifies formal characteristics of their own Typical.

| 296 | R | That one *[takes Typical Number Report Sheet]* I find like – because 19 is somehow *[points at dots around* | Den hier find ich so *[nimmt Typische Zahl Steckbrief]* weil 19 ist irgendwie *[deutet auf die Punkte, die bei ca.* |

		-19°C] – well yes there is a lot, but – I'd take like such an area *[draws a line on x-Axis from -25°C to -8°C]*, take the area here and then there is the middle like here *[points to around -16°C]* at, where are we, at 16 roughly the middle *[writes 16 onto diagram]* – which is why I'd take 16.	-19° Grad liegen], ja da ist auch schon viele, aber – ich würd irgend-wie so einen Bereich auch *[zeichnet einen Strich unter die x-Achse bei - 25° bis -8° Grad]* den Bereich in diesem nehmen und dann ist ja eher hier so die Mitte *[deutet auf ca. -16° Grad]* bei, wo sind wir, bei 16 unge-fähr die Mitte *[schreibt eine 16 ins Diagramm]* – deswegen würd ich hier so 16 nehmen.
297	DL	#I didn't quite understand this.	#Das hab ich jetzt nicht ganz verstan-den.
298	Q	#*[incomprehensible]*	#*[Unverständlich]*
299	R	Well because here there are the most dots, in that area *[points at area of the line drawn from -25°C to -8°C]*.	Ja weil hier sind so die meisten Punkte in dem Bereich *[deutet den Bereich der Linie von -25° bis -8° Grad an]*.
300	DL	Um.	Mhm.
301	R	And, uh, then I'd pick roughly the middle from that *[points to around - 16°C]* because there are an approxi-mately equal amount of dots every-where *[again points to same area]*, because of that roughly the middle *[again points to around -16°C]*, that's why I'd pick 16.	Und, ähm, dann würd ich ungefähr die Mitte davon nehmen *[deutet auf ca. -16°C]*, weil da sind überall ungefähr so gleich viele Punkte *[deutet den Bereich erneut an]*, deswegen ungefähr die Mitte *[deutet nochmal ungefähr -16° Grad an]*, deswegen würd ich um die 16 so nehmen.
302	DL	And what then is 16?	Und was ist die 16 dann?
303	R	16 degrees then is roughly typical.	16 Grad ist ungefähr so typisch dann.

Analysis of mathematizing side. Comparing their own use of $||MCs/m\text{: } Typical||$ to the Number Report Sheet, Rebecca disagrees with the value of 19 (#296). Instead, she engages in formalizing and representing to explain the correct value. From her activities, the theorem-in-action $<MA/r\text{: } Typical \text{ } represents \text{ } the \text{ } area \text{ } of \text{ } most \text{ } dots>$ ("here there are the most dots in that area", #299) can be reconstructed. Rebecca even gives a method of calculation for $||MCs/m\text{: } Typical||$: her method follows the theorem-in-action that $<MA/f\text{: } to \text{ } find \text{ } Typical, \text{ } one \text{ } needs \text{ } to \text{ } find \text{ } the \text{ } area \text{ } of \text{ } most \text{ } dots \text{ } and \text{ } then \text{ } take \text{ } the \text{ } middle \text{ } value \text{ } of \text{ } that \text{ } area>$ (#299, #301, #303). Although the value of Typical then is a single number, Typical represents the whole $||MCs/f\text{: } area \text{ } of \text{ } most \text{ } dots||$, because in this area there are $||MCs/f\text{: } equally \text{ } many \text{ } dots||$ ("pick roughly the middle from that because there are an approximately equal amount of dots everywhere", #301) – possibly meaning that in this area (-25 to -8) the data show almost uniform frequencies, and can thus be represented by a single value. Thus, the situative feature of $||MCs/f\text{: } equally \text{ } many \text{ } dots||$ could be a precursor to the general feature of an $||MCg/f\text{: } uniform \text{ } distribution||$. Thus, a formal characteristic of Typical is being the $||MCs/c:$ *middle of an interval*$||$. It, however, remains unclear how to find the suitable $||MCs/f:$ *area of most dots*$||$.

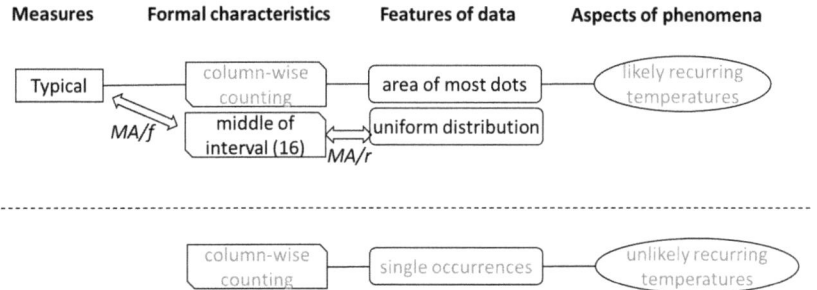

Fig. 7.30: Quanna and Rebecca engage in formalizing and representing to develop their own measure of Typical

Effects of design elements for the Typical Antarctic Temperatures Problem

Compared to Maria and Natalie's learning process during the Typical Antarctic Temperatures Problem (Fig. 7.18), Quanna and Rebecca's conceptual network remains comparatively less dense (Fig. 7.31). The students begin by engaging in structuring, identifying features of the data and aspects of the phenomenon, as well as linking the different concepts. Afterwards, no further structuring activity takes place, and the students mostly do not broaden the variety of their aspects of the phenomenon or features of the data. This does not mean that their learning process concerning the mathematizing side exhibits a lesser quality, as the students do manage the problems presented to them. Their activities only seem more focused. The pair even arrives at a development of Typical that includes formal characteristics as well as features of the data and aspects of the phenomenon.

Again, various effects of the design elements could be observed (Table 7.24). In sharp contrast to the processes of Maria and Natalie, Quanna and Rebecca do not engage much with the phenomenon of Antarctic temperatures. In the earlier phases, they only identify two aspects of the phenomenon; later, they do not further engage in structuring the phenomenon. Although the two aspects of the phenomenon identified show similarities to the intended concepts-in-action, this whole de-emphasis on structuring overall was not anticipated. Possibly, the design experiment leader would have needed to intervene, but did not give the required prompts. This shows how the use of a context alone does not necessarily lead to structuring activity, but that the rest of the task design (including behavior of the design experiment leader) also has to stimulate such activities. Overall, the design principle DP/Context thus seems to have had little effect on the students' learning processes.

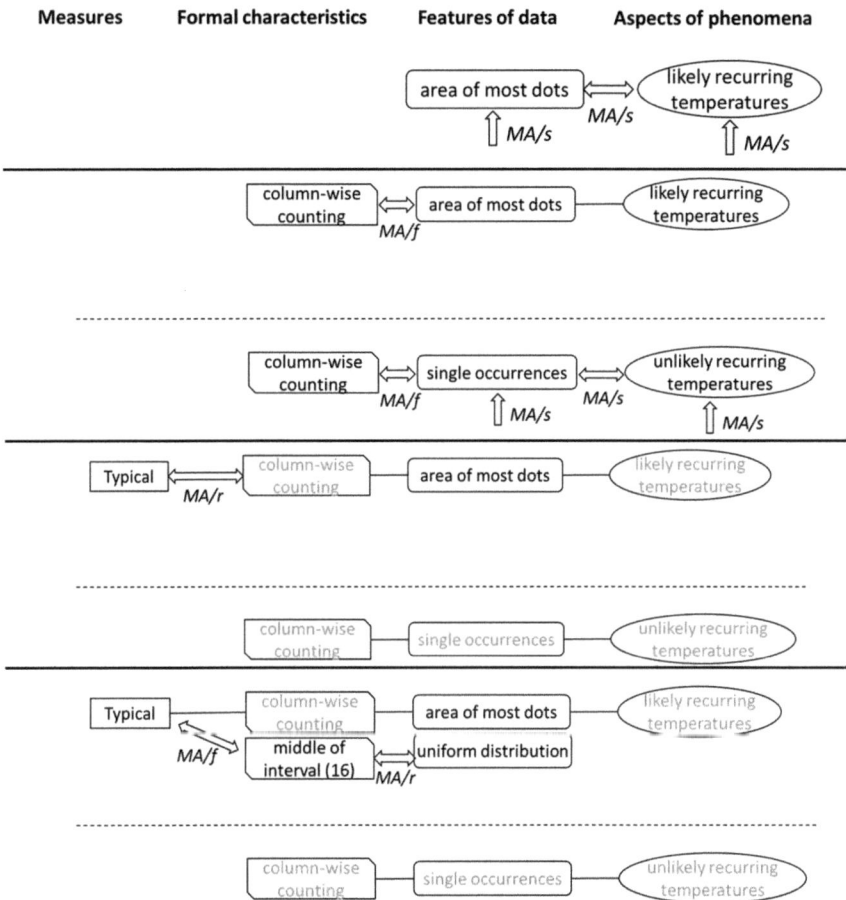

Fig. 7.31: Overview on Quanna and Rebeccas learning process during the Typical Antarctic Temperatures Problem

Tab. 7.25: Intended and actual concepts-in-action stimulated through the design elements of the Typical Antarctic Temperatures Problem. Printed in bold are unanticipated concepts-in-action.

Design element	Intended concepts-in-action	Actual concepts-in-action
Antarctic weather context	*||MCs/a: short-term variability||* *||MCs/a: long-term patterns of center and spread||* *||MCs/a: possibility of predicting temperatures||* *||MCs/a: typical temperatures||* *||MCs/a: exceptional days||*	*||MCs/a: likely recurring temperatures||* *||MCs/a: unlikely recurring temperatures||*
Single-year data	*||MCs/f: Two modes||* *||MCs/f: Range from -23.5 to -4.5||* *||MCs/f: Middle 50% from -17 to -9||*	*||MCs/f: area of most dots||*
Three-year data		***||MCs/c: column-wise counting||***
	||MCs/f: Different modes|| *||MCs/f: Range from -35 to -4||* *||MCs/f: Similar Middle 50% for 2002 and 2004||*	*||MCs/f: area of most dots||* *||MCs/f: single occurrences||*
Empty report sheet	*||MCg/m: Minimum||* *||MCg/m: Maximum||* *||MCs/m: Typical||* *||MCg/c: Method of calculation||* *||MCg/c: Interval||* *||MCs/a: typical temperatures||*	*||MCg/m: Minimum||* *||MCg/m: Maximum||* *||MCs/m: Typical||* *||MCg/c: number||*
Filled-in report sheets	*||MCg/c: Interval||* *||MCg/c: Number||* *||MCs/c: Single highest frequency||* *||MCs/f: Main body of data||* *||MCs/f: Area encompassing most modes||* *||MCs/f: Highest density in 2002 and 2004||* *||MCg/f: Boundary of data||* *||MCg/f: Mode of 2003||* *||MCg/a: Worst observed case||* *||MCs/a: Exceptional days||* *||MCs/a: Most likely temperatures||* *||MCs/a: Reasonable possible temperatures||* *||MCs/a: Reasonably expected temperatures||*	***||MCs/c: middle of an interval||*** ***||MCs/f: equally many dots||*** ***(no identified aspects)***

Overall, Quanna and Rebecca's process did not show the variety of concepts drawn on by Maria and Natalie, but a rather more focused and efficient use of concepts. Instead of a focus on structuring, the students placed a focus on formalizing: without much input by the design experiment leader, they identified formal characteristics like the methods of calculation of ||*MCs/c: column-wise counting*|| and finding the ||*MCs/c: middle of an interval*||. In that, the design principle DP/Formalizations seems to have an increased effect on Quanna and Rebecca's activities than with Maria and Natalie.

Arctic Sea Ice Problem – Episode 1

In their learning process during the Arctic Sea Ice Problem, Quanna and Rebecca develop Typical during the creation of their report sheet in Phase 4. Previously, they did not much focus on Typical. They recognized the differences in summaries of the filled-in report sheets (Fig. 7.8), but had trouble articulating their own perspective: "It sort of changed actually – but not like, really" (Rebecca in #128, not printed). However, they start creating their own report sheet much earlier than Maria and Natalie, as their evaluation of the filled-in report sheet did not go into as much detail.

Shortly after progressing to Phase 4, Quanna and Rebecca begin creating their own report sheet (Fig. 7.32) by first intuitively finding values for Typical (with no discernible method of calculation). This analysis starts with the students turning towards the summary; the sketch gets produced only afterwards.

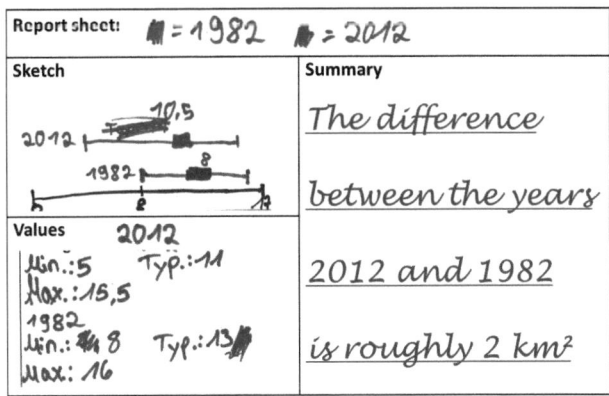

Fig. 7.32: Quanna and Rebecca's report sheet for Arctic sea ice

III-2b-QR; Phase 4; Start: 23:30
Previously, the students recognized the different summaries of the filled-in report sheets, but did not find an explanation. Now, the students fill out their own report sheet. While drawing a summary, they connect Typical to an aspect of the phenomenon.

194	Q	Right. - Now summary.	So. – Nun Zusammenfassung.
195	R	Um – there – the numbers got *[points to own report sheet]* – always – look – more and more ice was melting.	Ähh – da – die Zahlen wurden *[deutet auf den Steckbrief]* – immer – guck – da schmolz immer mehr Eis weg.
196	Q	*[shakes head]* the difference is – by 2.5.	*[Schüttelt mit dem Kopf]* der Unterschied ist – geht um 2,5.
197	R	Always?	Immer?
198	Q	Yes here *[points to written summary on report sheet]* of typical.	Ja hier *[deutet auf das Aufgeschriebene auf dem Steckbrief]* – von typisch.
199	R	Is that – varies by around 2? Km².	Ist das – variiert um ca. 2? Km²
200	Q	All differences.	Jeder Unterschied.
201	R	Yeah look, it is. Between here and there *[points to report sheet]* it's 3.	Ja ist doch. Zwischen da und da *[deutet auf den Steckbrief]* sind 3.
202	Q	11 and 13.5 is 2.5.	#11 und 13,5 ist 2,5.
203	R	Yes and here *[again points to report sheet]* is, ähm,– is 1, nah 0.5 there is 3. Therefore 2. – is roughly the middle.	Ja und hier *[deutet erneut auf den Steckbrief]* ist äh – ist 1, ne 0, 5 da ist 3. Deswegen 2. – Ist ungefähr die Mitte.
204	Q	Okay. You mean typical, right?	Okay. Du meinst typisch ne?
205	DL	What difference are you guys talking about?	Über welchem Unterschied sprecht ihr gerade?
206	R	The difference between 2012 and 1982 *[points to report sheet]*.	Den Unterschied zwischen 2012 und 1982 *[deutet das auf dem Steckbrief an]*.
207	Q	But then you need to take what's typical, because that's like approximately.	Aber da muss man ja typisch nehmen, weil das ist ja so ungefähr.

Analysis of mathematizing side. The students try to find a way to describe the $||MCs/a$: *increase in ice melt*$||$ that happened between 1982 and 2012 (that "ever more ice melted", #195). Quanna finds that $||MCs/m$: *Typical*$||$ can be a fitting measure to represent this change (the difference of Typical, #196, #198). For both years, the students already intuitively determined the values of 11 and 13.5 for Typical (see their own report sheet, Fig. 33, later changed to 13), so that $<MA/r$: *the difference of Typical by 2.5 represents the increase in ice melt*$>$ (#202).

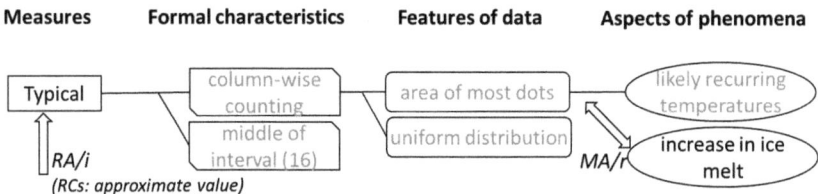

Fig. 7.33: Quanna and Rebecca continue their development of Typical by engaging in representing. They also identify a pattern of thought for Typical

Analysis of reflective activities. To justify her choice of $||MCs/m:$ *Typical*$||$ for representing the $||MCs/a:$ *increase in ice melt*$||$, Quanna states that "you have to use typical, because that is, like, roughly" (#207). In this way, Quanna does not justify the use of Typical because of a relation to a represented feature of the data or aspect of the phenomenon. This can be interpreted as Quanna engaging in the reflective activity of identifying patterns of thought with the underlying reflective theorem-in-action <*RA/i: Typical is an approximate value*>.

Analysis of reflective concepts. It is not entirely clear what Quanna means by drawing on her concept-in-action of $||RCs:$ *approximate value*$||$. This concept-in-action could be a situative precursor to the general reflective concept of $||RCg:$ *summary*$||$, the fact that measures reduce complex phenomena to numbers and thus allow further numerical treatment (such as calculating a difference). However, as Quanna does not explain further, this can count only as a tentative interpretation.

Arctic Sea Ice Problem – Episode 2

Some minutes later, the students finish their report sheet. During the previous design experiment, the pair had used Typical as if it was the middle of a typical interval. Now they used single numbers as formal characteristics of Typical in their report sheet. The design experiment leader notices this and asks the students to explain.

III-2b-QR; Phase 4; Start: 32:00
Previously, Quanna and Rebecca finished creating their own report sheet. Now, while explaining their use of Typical on their own report sheet, the students show conflicting views of the formal characteristics of Typical. They solve this conflict by identifying a corresponding feature of the data.

311	DL	Looks like you came to the decision of using a single value as typical. On the other hand, here *[points to Typical Interval Report Sheet]* here are – we are given a range.	Ich seh auch, ihr habt euch entschieden, typisch nur einen Wert zu nehmen. Im Gegensatz zu hier *[deutet auf den Typischer Intervall Steckbrief]* hier sind – ist immer so ein Bereich angegeben.

312	R	Um.	Mhm.
313	DL	Does it mean it's better or worse, what do you think?	Ist das besser oder schlechter, oder wie meint ihr das?
314	R	Well typical really is more like #a single, like…	Ja typisch ist ja eigentlich eher #so eine, also…
315	Q	#more like an area.	#mehr so ein Bereich.
316	R	Well – we are kind of contradicting ourselves but – typical –, ähm,it is – well not like, this here *[points to Typical Interval Report Sheet]* is the entire area, almost.	Also ja – irgendwie wiedersprechen wir uns gerade selber, aber – typisch –, ähm,ist ja – jetzt nicht, der *[deutet auf Steckbrief 2]* zeigt jetzt den ganzen Bereich angegeben fast.
317	DL	Um.	Mhm.
318	R	Yet typical is more like - either a small area or presumably some number – you could prefix it as a range – like this, for example, #you write like 10 to 12.	Und typisch ist ja eher so – entweder ein kleiner Bereich oder man kann auch vermutlich ne Zahl sagen also – man könnte jetzt auch einen Bereich davor schreiben – zum Beispiel hier so #10 bis 12 oder so aufschreiben.
319	Q	#typical *[clears throat]* typical always like the middle.	#typisch *[räuspert sich]* typisch immer so die Mitte so.
320	R	#Yes.	#Ja
321	DL	#Um.	#Mhm.
322	R	Or here, where there's more - if like, if we would like, uh -	Oder da, wo halt mehr – wenn jetzt, wenn wir jetzt ääh -
323	Q	There is lots of dots bunched up.	Viele Punkte auf einem Fleck sind.

Analysis of mathematizing side. Immediately, the different formalizations of Typical are problematized as the students seem to hold different views: $||MCs/m: Typical||$ could take the form of a $||MCg/c: Number||$ (#314) or an $||MCg/m: Interval||$ (#315). Rebecca starts to consolidate these two different formal characteristics by contrasting their own Typical to that on the Typical Interval Report Sheet, which uses a very large interval (#316). To clarify, the pair engages in the mathematizing activity of representing a feature already identified in the first design experiment: <*MA/r: Typical represents the area of most dots*> (#322, where there are "lots of dots bunched up", #323). The actual form of Typical does not seem to be much of a concern: <*MA/f: Typical can take the form of an interval or the middle of an interval*> (#318, #319).

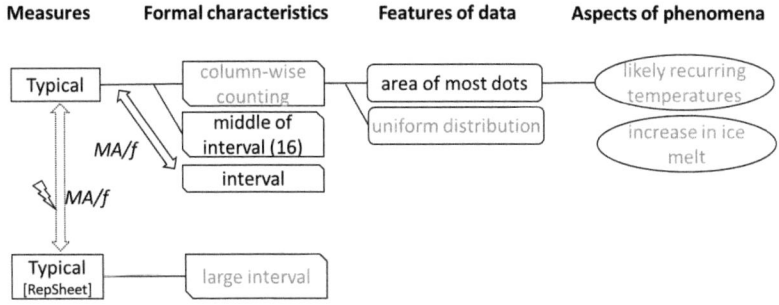

Fig. 7.34: Quanna and Rebecca contrast their own measure of Typical what that on the Large Interval Report Sheet, which prompts them to further formalize their own measure

Arctic Sea Ice Problem – Episode 3

A short time later, the design experiment leader again challenges the students to commit to whether Typical has the formal characteristic of an interval or a number.

III-2b-QR; Phase 4; Start: 33:30
Previously, Quanna and Rebecca engaged in formalizing, but did not clarify whether Typical is an interval or a number. Now, the design experiment leader prompts the students to explain. Rebecca identifies another formal characteristic.

332	R	Typical is actually a small range, so I think, typical shouldn't be that big now, because - typical, you could express everything using typical and so - typical is between 2.5 and 16.	Typisch ist eigentlich auch ein kleiner Bereich, aber ich finde, typisch sollte jetzt nicht so groß sein, weil - typisch, da kommt man auch einfach alles in typisch angeben und jetzt – typisch ist 2,5 bis 16.
333	DL	Um.	Mhm.
334	R	That would be – *[whispering to Quanna]*	Das wär - *[flüstert an Quanna]*
335	Q	Uh - nothing, really.	Ähm - eigentlich nichts.
336	R	Okay.	Okay.
337	DL	[...] Got it! Um. You just said that it's stupid when typical is too large, but even so somehow - what would count as too big or too small? How about it?	[...] Alles klar. – Mhh – jetzt hast du grad erzählt, wenn typisch zu groß ist, ist doof, aber irgendwie auch wenn – was wär denn zu groß oder zu klein? Wie soll das denn so?
338	R	Well – it should be, like – it depends on how big of an area it is.	Ja – es sollte jetzt – kommt drauf an, wie groß der Bereich ist.
339	DL	Um.	Mhm.
340	R	If the area is larger than 100 or something, then it could go higher than 10 – but if the area only was like, um, 10.5, then it should not get over 5 or something, um, under no circumstance higher than half – a bit like that – not actually at 10.5 or 10, I'd say like 2.3 to 4 – 4.	Wenn der Bereich jetzt über 100 ist oder so, dann darf das schon mal über 10 sein – aber wenn der Bereich jetzt nur so, ähm,10,5 ist, dann sollte er nicht über 5 kommen oder so mh auf keinen Fall mehr als die Hälfte – irgendwie so – dann jetzt bei 10,5 |oder 10, würd ich so sagen so 2,3 bis 4 – 4.

Analysis of mathematizing side. In this scene, the students again engage in formalizing by providing further formal characteristics for Typical. Rebecca acknowledges that <*MA/f: Typical takes the form of an interval*>, which should not be too big ("typical shouldn't be that big", #332). Pressed on the issue of how big Typical should be, Rebecca approximates the extent of Typical through several examples, engaging in formalizing that <*MA/f: Typical should not cover more than half the data*> (#340).

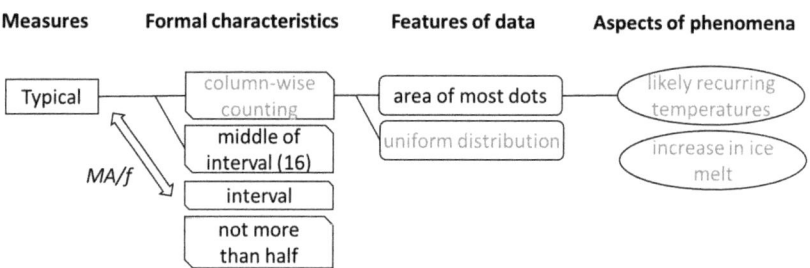

Fig. 7.35: Quanna and Rebecca further engage in formalizing Typical by identifying additional formal characteristics

Effects of design elements for the Arctic Sea Ice Problem

During the Arctic Sea Ice Problem, Quanna and Rebecca mostly focus on the measure of Typical (Fig. 7.36). Their focus on formalizing that already could be observed during the Typical Antarctic Temperatures Problem becomes even more apparent, as the students mostly focus on identifying formal characteristics of Typical. Only rarely do they refer to any aspects of the phenomenon.

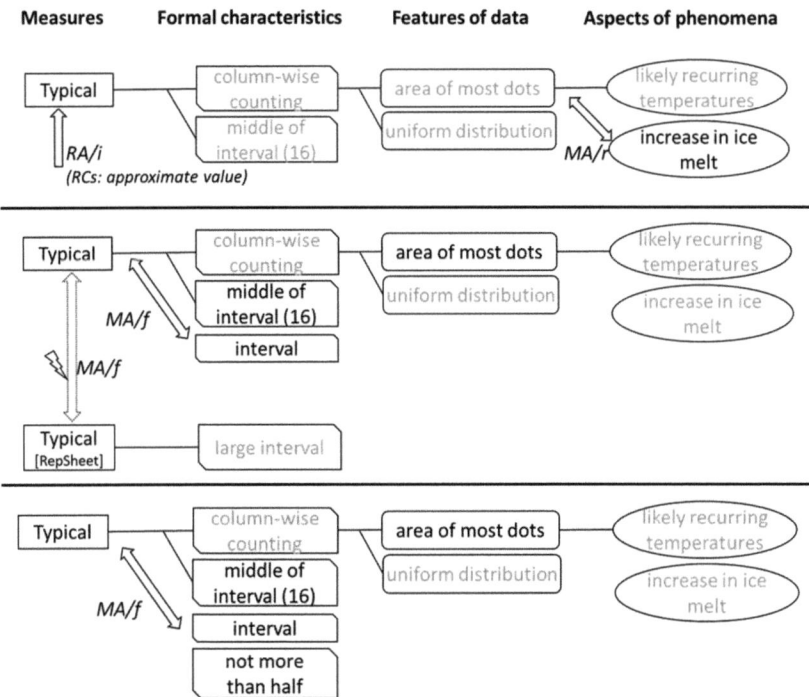

Fig. 7.36: Overview on Quanna and Rebecca's learning process during the Arctic Sea Ice Problem

As with the effects for the Typical Antarctic Temperatures Problem, the effects of the design elements show a focus on eliciting the mathematizing activity of formalizing (Table 7.26). In their formalizing activities, the students even unexpectedly identified a formal characteristic of Typical that shows similarities to the general measure interquartile range: Typical should never encompass more than half of the data, whereas the interquartile range encompasses exactly half of the data. The design principle DP/Formalization, which aimed at eliciting formalizing by encouraging commitment to values and by providing different possible formalizations, thus again seems to have greatly influenced the students' learning processes.

Tab. 7.26: Intended and actual concepts-in-action stimulated through the design elements of the Arctic Sea Ice Problem. Printed in bold unanticipated concepts-in-action.

Design element	Intended concepts-in-action	Actual concepts-in-action
Arctic sea ice context	*||MCs/a: winter months||* *||MCs/a: summer months||* *||MCs/a: melting process||*	
Two-year data	*||MCs/f: modes||* *||MCs/f: left-skewed||* *||MCs/f: spread from 7.5 to 16.5||*	
Three-year data	*||MCs/f: increased spread in data||* *||MCs/a: increasing severity of ice melt||*	
Empty report sheet	*||MCg/m: range||*	***||MCs/m: Typical||*** *||MCs/a: increase in ice melt||*
Filled-in report sheets	*||MCg/m: range||* *||MCs/m: Typical||* *||MCg/c: number||* *||MCg/c: interval||* *||MCg/c: calculation through min. and max.||* *||MCs/f: main body of data||* *||MCg/f: spread of data||* *||MCs/a: beginning of winter||* *||MCs/a: severity of ice melt||*	*||MCs/m: Typical||* *||MCs/c: interval||* ***||MCs/c: not more than half||*** *||Mcs/f: area of most dots||*

DP/Context, which aimed at eliciting structuring through providing a suitable context, played a smaller role, as the aspect of *\|\|MCs/a: increase in ice melt\|\|* was identified, but did not have much further influence on the students' activities. As with the learning process of Maria and Natalie, DP/Measures also took little effect, as the students used Typical – and not the better fitting range – to represent the increase in ice melt.

7.2.4 Overarching effects for Quanna and Rebecca's learning processes

Connecting aspects and features

In contrast to Maria and Natalie who first focus on identifying features, Quanna and Rebecca start out by identifying the aspect of *\|\|MCs/a: Likely recurring temperatures\|\|* and connecting it to the feature of *\|\|MCs/f: Area of most dots\|\|*. This seems to provide the students with a sufficiently firm understanding of the phenomenon, so that other activities of structuring do not seem necessary to them. They also are quick to represent this aspect through the measure $\|MCs/m:$ *Typical*$\|$. Other than identifying the complementary aspect $\|MCs/a:$ *Unlikely recurring temperatures*$\|$, no new aspects need to be represented. On the one hand, this presents an extremely efficient way to deal with the problem. On the other hand, the resulting conceptual network does not show the richness observed with Maria and Natalie. Correspondingly, they show little representing activity during the Typical Antarctic Temperatures problem, as there are no further identified aspects of the phenomenon to be represented by measures (Fig. 7.37).

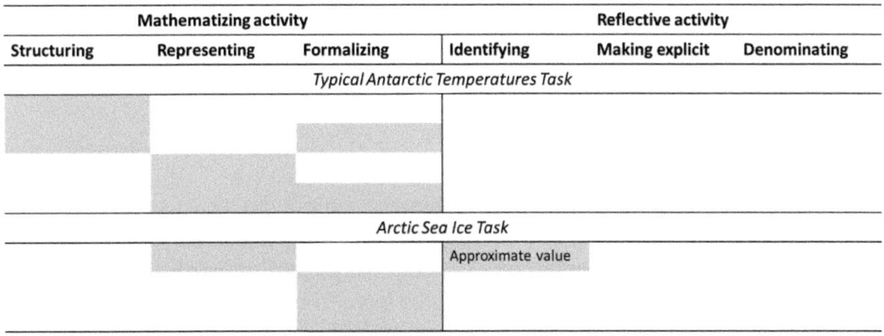

Mathematizing activity			Reflective activity		
Structuring	Representing	Formalizing	Identifying	Making explicit	Denominating
Typical Antarctic Temperatures Task					
Arctic Sea Ice Task					
			Approximate value		

Fig. 7.37: Quanna and Rebecca's mathematizing and reflective activities. Highlighted in grey are the pursued activities in each scene. For the reflective side, the reflective concepts-in-action are given.

Emphasizing formalizing

Fig. 7.37 also reveals that, whereas Maria and Natalie showed an emphasis on structuring, Quanna and Rebecca show an emphasis on formalizing. This shows how the design principle of DP/Context was less important to their learning process than the design principle DP/Formalizing. Thus, in the end, their conceptual network of measures, aspects, and features is much smaller than that of Maria and Natalie, but shows more detail in formal characteristics.

Mathematizing without reflecting

Although their learning pathways are different, Quanna and Rebecca develop a measure $\|MCs:$ *Typical*$\|$ similar to that of Maria and Natalie: representing the $\|$*likely recurring temperatures*$\|$, corresponding to the feature of the $\|MCs:$ *Area of most dots*$\|$, taking the form of an $\|MCg:$ *Interval*$\|$. Yet whereas Maria and Natalie also show various reflective activities, Quanna and Rebecca do not (Fig. 7.37). A possible explanation could be the decreased emphasis on structuring, since the analysis of Maria and Natalie revealed the importance of context for reflection. This does not mean that Quanna and Rebecca's mathematizing pathways are lacking in quality; however, they had only limited learning opportunity for developing the reflective side of mathematical literacy in statistics. Reflection thus does not seem an activity automatically elicited by the teaching-learning arrangement, but instead needs additional focus for further design.

7.3 Students' development of measures in Cycle III

This Chapter tracked the development of mathematizing concepts of two pairs of students on a micro level in order to find connections between students' situative mathematizing concepts and general mathematizing concepts. Comparing the learning processes of Maria and Natalie with those of Quanna and Rebecca yields some insights into how this design of a teaching-learning arrangement can support the development of mathematical literacy in statistics.

7.3.1 Connections to general measures

The aim of a teaching-learning arrangement cannot be the use of situative measures alone. Instead, the use of situative measures needs to support the development of general mathematizing concepts of general measures. For that, the connections of situative to general measures have to be explicated.

Tab. 7.27: Maria and Natalie's (M&N) and Quanna and Rebecca's (Q&R) respective situative measures 'Typical' show similarities to the general measure 'interquartile range'

Measure	Characteristics	Features	Aspects
Typical (M&N)	Interval, Actual data, Calculation through average	Area of most dots	How it is most of the time, Always varying, Individual days
Typical (Q&R)	Interval, Not more than half, Middle of interval	Area of most dots	Likely recurring temperatures, Increased ice melt
Interquartile range	Length of interval, Middle 50% of data, Calculation through median	Main area of data	Expected deviation from expectation

Table 7.27 reveals the similarities between the students' situative measures Typical and the general measure interquartile range. The two pairs' developments of Typical show formal characteristics, features of data, and aspects of phenomena that either already are almost the same (not more than half data – middle 50% of data; calculation through average – calculation through mean) or can be seen as situative variants of abstract aspects of phenomena (how it is most of the time / likely recurring temperatures – expected deviation from expectation).

The students thus developed situative precursors to general concepts through their development of situative measures. This shows the potential of developing general measures from students' situative measures. This last step, however, was missing in the design experiments, as it was not yet designed for in Cycle III.

7.3.2 The broadening of the contextual neighborhood

One part of the development of general measures, however, could already be observed. Chapter 4 showed how the development of general mathematizing concept can be described as a process of broadening a measure's contextual neighborhood. During the Arctic Sea Ice Problem, Maria and Natalie use language related to the phenomenon of Antarctic temperatures to talk about Arctic sea ice and to reflect on their measure Typical. Although at first this seems to happen by accident, the students later remark on their use of language, yet continue to frame discussion of Typical in terms of temperatures. Similarly, when formalizing Typical for Arctic sea ice, Quanna and Rebecca re-discover the feature of the area of most dots they already identified for Antarctic temperatures. For both pairs, certain aspects of phenomena and features of the data seem to 'carry over' from one design experiment to the other – they are in process of broadening the contextual neighborhood of their situated abstraction of Typical.

7.3.3 Differences in richness of activity

Although both pairs develop a promising situative measure Typical, the learning processes show wildly different richness in activity. Whereas Maria and Natalie engage fully in all mathematizing activities and often multiple activities simultaneously, Quanna and Rebecca mostly focus on formalizing and eschew the activity of structuring. And whereas for Maria and Natalie these mathematizing activities are often accompanied by reflective activities, Quanna and Rebecca seldom engage in reflective activity. This results in a mathematizing conceptual network that is much broader for Maria and Natalie than for Quanna and Rebecca, and in much more opportunities for Maria and Natalie to develop reflective concepts.

Two diverging points in the learning processes could provide possible explanations for these differences. Maria and Natalie (1) strongly engage in the mathematizing activity of structuring, elicited by the design principle DP/Context. This provides them a language for reflection, and the identification of additional aspects enables to contrast measures regarding their aims and purposes and risks and limits. Maria and Natalie also (2) contrast Typical with the average, prompting additional reflective activity to explain the differences. Such a contrasting of measures was intended to be initiated by the design principle DP/Measures through the filled-in report sheets. However, it did not become important for the development of Typical; Maria and Natalie's familiarity with the average filled that role.

7.3.4 Unforeseen effects of design principles

One goal of the analysis was to reveal the effects of the design principles on the learning processes. Some of those effects were unforeseen. From those unforeseen effects, some proved beneficial, such as the familiar context providing a language for reflection. Some effects proved to be obstacles as well as beneficial, like the identification of aspects of the phenomenon that had no corresponding features in the data.

Other effects steered the learning processes into new directions. One of those is the unexpected focus on Typical for the Arctic Sea Ice Problem. It was intended by design that Typical would be understood as a measure that proved not much use for Arctic sea ice, as the declining minimum or the increased range would have much higher contextual relevance than a typical size of sea ice. The design principle DP/Measures sometimes initiated the reflection that the perspectivity of Typical obscures the truly important aspect of Arctic ice melt. Although Maria and Natalie did denominate risks and limits by drawing on precursor situative concepts of perspectivity, this insight was not induced by the design.

Instead, both pairs further engaged in formalizing Typical rather than use other measures. One explanation could be the prominent place of Typical on the report sheets for Antarctic temperatures, prompting the students to repeat the use of this situative measure. Cycle IV (in the following chapter) builds on this insight to explore options for initiating more reflective activity.

8 Students' reflective concepts

The analysis of Cycle III in Chapter 7 revealed how students can develop their situative measure of Typical by engaging in mathematizing activities and broadening the contextual neighborhood of their situated abstractions. The resulting situative measures showed some similarity to the general measures. It also appeared that the students' reflective activities had an impact on their mathematizing activities, and that comparing different measures seems to be a task that elicits both types of activity. The students' reflective activities and concepts, however, played only a minor role in the analysis of Chapter 7. To provide empirical insights into the development of the reflective side of mathematical literacy in statistics (research question RQ2), this chapter focuses on the initiation of reflective activities and the identification of reflective concepts.

8.1 Design of a teaching-learning arrangement for Cycle IV

The design of the teaching-learning arrangement for Cycle IV was mostly built upon that of Cycle III with a few adaptations. Since the comparison of different measures proved so helpful in the Cycle III, yet was not engaged by every pair of students, the teaching-learning arrangement in Cycle IV strengthened the emphasis on the comparison of different measures. In Cycle IV, the design experiment series consisted of three design experiment sessions, working with three problems: the Antarctic Temperatures Problem (slightly modified from Cycle III, Section 8.1.2), the Arctic Sea Ice Problem (also slightly modified from Cycle III, Section 8.1.3), and the Context Comparison Problem (Section 8.1.4 and Chapter 9).

8.1.1 Refined design principles for initiating reflective activities

Because this cycle focuses on the initiation of reflective activity, the design principles have to be chosen accordingly. Cycle III already showed how some of the design principles such as DP/Context, which intended to initiate mathematizing activity, also supported the students' reflective activity. For Cycle IV, some of the design principles were refined for reflective activities based on the experiences with the learning processes of Cycle III, whereas others were simply carried over.

Drawing on the epistemic and reflection-initiating functions of context (DP/Context-Refined)

The analysis in Chapter 7 revealed how the context played an important part not only for mathematizing concepts and activities: The context allowed the students to identify aspects of phenomena, thus supporting mathematizing activi-

© Springer Fachmedien Wiesbaden GmbH, part of Springer Nature 2018
C. Büscher, *Mathematical Literacy on Statistical Measures*,
Dortmunder Beiträge zur Entwicklung und Erforschung des
Mathematikunterrichts 37, https://doi.org/10.1007/978-3-658-23069-2_8

ties; it also allowed the students to evaluate the filled-in report sheets, to denominate risks regarding reliance on ill-fitting measures for Antarctic temperatures, thus also supporting reflective activities. The context was not only a possible object of reflection (as in the context-oriented reflection specified by Skovsmose, 1998), but also supported the reflection on the use of statistical measures. The context thus did not only have an *epistemic* function, but also a *reflection-initiating* function. A teaching-learning arrangement aiming for mathematical literacy in statistics thus needs to choose a context suitable for this reflection-initiating function: *To enable concept-development based on learners' situative mathematizing **and reflective** concepts, choose a phenomenon familiar to the students for the context of the problem, because this can enable them to engage in the mathematizing activity of structuring **as well as in various reflective activities** ('DP/Context-**Refined***').

Contrasting different measures (DP/Measures-Refined)

Chapter 7 also revealed that the design principle DP/Measures, however, fell short of expectations. This design principle was introduced in order to enable students' to consciously choose between different measures, creating reflective situations of advocating or rejecting measures. Instead, the students mostly focused on using the single situative measure of Typical.

However, when Maria and Natalie contrasted the measures average and Typical, the effects intended by DP/Measures did emerge. The pair engaged in rich mathematizing activity in explaining which aspects of the phenomenon were represented by which measure. Additionally, reflective situations were created, and reflective activities were elicited, as the students also compared the two measures through the reflective concept-in-action of $||RCs$: what it tells you$||$, a situative precursor to $||RCg$: perspectivity$||$. Explicitly comparing and contrasting students' processes and products is an activity that thus seems to support reflective activities (see also Lengnink, 2013). Contrasting of measures also elicited mathematizing as well as reflective activities and the development of mathematizing and reflective concepts. This insight presents a reason to refine the design principle DP/Measures: *To increase the conceptual richness of learners' mathematizing **and reflective** activities, put them into a position that **requires** them to **contrast** different measures, because **contrasting measures** can prompt further mathematizing activities of structuring and representing **as well as reflective activities of explicating aims and purposes and denominating risks and limits** ('DP/Measures-**Refined***').

8.1.2 The Antarctic Temperatures problem

Data and setting of this problem are adopted from the Typical Antarctic Temperatures problem from Cycle III. However, the focus on Typical is dropped in favor of contrasting a multitude of different situative measures (implementing DP/Measures-Refined). This necessitates a change in the filled-in report sheets and the sequence of phases. Table 8.1 gives a short overview of the phases of the Antarctic Temperatures problem.

Tab. 8.1: The phases of the Antarctic Temperatures Problem

#	Name	Description
1	Introducing the context	Introduction to the role of advisors to Antarctic researchers (same as in Cycle III)
2	Predicting phenomena	Predicting ten days for July 2015 based on data from July 2002 (same as in Cycle III)
3	Dealing with year-to-year variability	Changing the prediction to account for data from 2003 and 2004 (same as in Cycle III)
4	Comparing report sheets	Contrasting filled-in report sheets using different situative measures
5	Creating own report sheet	Creating own report sheet with a focus on choosing a measure

As can be seen, the phases regarding predicting the temperatures stay the same as in Cycle III. The sequence of Phases 4 and 5, however, is switched. This goes along with a change in the report sheets (Fig. 8.1, left side). The report sheets of the Antarctic Weather Problem no longer consist of the fixed measures of minimum, maximum, and Typical. Instead, they feature the same blank 'values'-field used in the Arctic Sea Ice Problem in the previous cycle. The measure Typical also now is used only on one of the filled-in report sheets. Thus, the measures minimum, maximum, and Typical are no longer guaranteed to be the object of discussion for the students. Since Typical now is no longer a fixed measure on the report sheets, the problem is now referred to as simply the Antarctic Weather Problem (foregoing the focus on Typical).

The report sheets were changed to provide a stronger emphasis on different measures. Each of the filled-in report sheets now uses a different situative measure, resulting in a different perspective on the phenomenon. The Typical Report Sheet is intended to introduce the idea of using an interval of values in the middle of the data to characterize the whole data set. In contrast, the Value Report Sheet uses the situative measure of Most Important Value to summarize the data only under a single value of center. The MinMax Report Sheet again summarizes the data by giving their maximal spread.

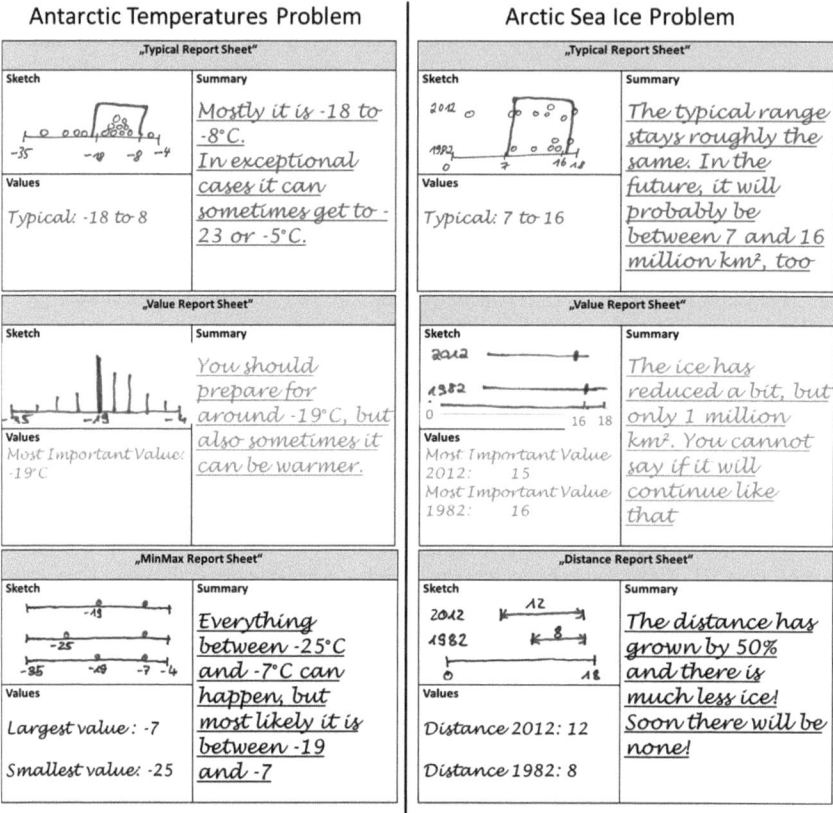

Fig. 8.1: The filled-in report sheets for Cycle IV

Each report sheet thus emphasizes a different perspective on the same phenomenon, sometimes taking the spread of the data into account, sometimes only the center, and sometimes a combination of both. When asked which report sheets is best, the students have to contrast the different measures and perspective. Thus, the filled-in report sheets realize one version of the design principle DP/Measures-Refined.

For creating their own report sheet, the students receive an empty report sheet featuring the same blank 'values' field, so that they are not forced to use Typical but instead can choose a fitting measure. However, because such an empty report sheet is not self-explanatory, the phases of the design experiment were switched, with the students receiving the filled-in report sheets in Phase 4, and creating their own report sheet in Phase 5. This allows the students to familiarize themselves with the idea of report sheets using the filled-in report sheets as examples. A consequence of this is that in creating their own report sheet, the students might be influenced by the filled-in report sheets, so that their own report sheet might no longer reflect their genuine intuitive reasoning. However, the aim of this design experiment cycle was not to identify intuitive reasoning concerning mathematizing concepts, but rather to investigate students' reflective activities when contrasting measures. The report sheets designed in this way support this aim by enabling discussion about a variety of different measures.

8.1.3 The Arctic Sea Ice Problem

The second design experiment is based on the Arctic Sea Ice Problem. This problem is mostly the same as in Cycle III, following the same progression of phases (Tab. 8.2). Minor changes were made to the filled-in report sheets (Fig. 8.1, right side). The filled-in report sheets of the Arctic Sea Ice Problem now correspond to the filled-in report sheets of the Antarctic Temperatures problem, using the same or similar situative measures and perspectives (indicated by the same color). Compared to Cycle III, however, the perspective introduced by each filled-in report sheet is made much clearer by providing summaries about the future of the Arctic sea ice: the Typical and Value Report Sheets argue that the ice will not change (much), whereas the Distance Report Sheet asserts that the ice is disappearing. A central aim of the design experiment is for the students to acknowledge that this difference in summaries lies not in simple manipulation of data, but in the different perspectives created by different measures – as such, the aim is for the students to reinvent precursors to the general reflective concepts of perspectivity.

Tab. 8.2: The phases of the Arctic Sea Ice Problem

#	Name	Description
1	Introducing the context	Introduction to the role of experts on Arctic sea ice and the Arctic sea ice data from 1982 and 1992 (same as in Cycle III)
2	Interpreting report sheets	Predicting Arctic sea ice in 2012 based on filled-in report sheets (same as in Cycle III, with minor changes to filled-in report sheets)
3	Evaluating report sheets	Evaluating filled-in report sheets on the basis of data from 1982, 1992, and 2012 (same as in Cycle III, with minor changes to filled-in report sheets)
4	Reporting on Arctic sea ice	Creating and justifying an own report sheet on Arctic sea ice

8.2 Empirical reconstruction of students' reflective concepts

The series of three problems was utilized in design experiments with three pairs of students in Cycle IV. This analysis follows two pairs of students, Kaan and Nesrin and Jana and Maria, during their learning processes. The aim of the analysis is to identify the students' situative reflective concepts and to show the possible general reflective concepts that could be developed from those. During their learning processes, the students engage in various reflective activities exhibiting a rich network of situative reflective concepts. Because of this richness, only the learning processes during the Arctic Sea Ice Problem are analyzed here; the processes during the Antarctic Temperatures problem showed similar reflective activities, but also along with a heavier focus on mathematizing activities. The method of analysis follows the methods introduced in Chapter 6 and applied in Chapter 7. The special attention to the effects of design elements and principles is dropped, however, in favor of an increased emphasis on the relation between situative and general reflective concepts.

This analysis expands on two articles in which preliminary and shorter versions have been presented (Büscher & Prediger, submitted; Büscher & Schnell, 2017; Büscher, submitted).

8.2.1 Kaan and Nesrin: Discovering the perspectivity of measures

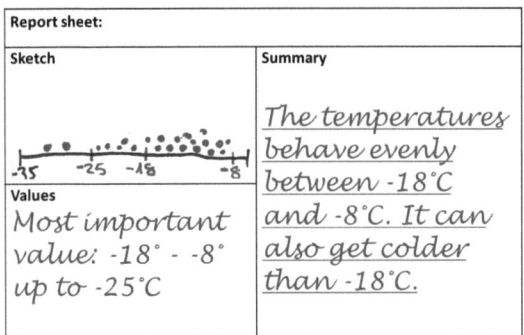

Fig. 8.2: Kaan and Nesrin's report sheet from the previous design experiment session concerning the Antarctic Temperatures Problem

The first sequence of the second design experiment session follows the students Kaan (K) and Nesrin (N) during the Arctic Sea Ice Problem. In the previous design experiment session concerning the Antarctic Temperatures Problem, the pair created a report sheet that combined elements of the measures Most Important Value and Typical (Fig. 8.2). Now, during the Arctic Sea Ice Problem, both students show high motivation in working on the problem, although Kaan seems a bit shy and leaves Nesrin to do much of the talking. In the course of the design experiment session, the pair initially has trouble interpreting the given filled-in report sheets, as the different assertions on the filled-in report sheets cannot be explained. By structuring the phenomenon and finding fitting representing measures, the students manage to overcome this difficulty and to reflect on the perspectivity of measures.

The analysis starts with Phase 2, as the students start to interpret the filled-in report sheets.

IV-2-KN; Phase 2; Start: 12:00
Previously, the students Kaan and Nesrin were introduced to the context of the design experiment. Now, they notice the different summaries of the Distance and Typical Report Sheets.

76	N	I don't get it. Here they say *[hints at Distance Report Sheet]* that it decreased here, but here *[Typical Report Sheet]* they say that it roughly remains the same.	Also ich versteh das nicht, wenn die *[deutet auf Abstand-Steckbrief]* sagen, dass das hier gesunken ist, aber hier *[deutet auf Typisch-Steckbrief]* sagen die, dass es ja ungefähr gleich bleibt.
77	K	Maybe they didn't measure correctly, like.	Vielleicht haben die nicht richtig gemessen oder so.
78	N	Well. I think that this *[Distance Report Sheet]* says that it increased by, by about half. But they *[Typical Report Sheet]* say that it's the same, it's roughly the same.	Ja. Ich find der *[deutet auf Abstand-Steckbrief]* sagt ja, dass die ca. um die, um die Hälfte gewachsen ist. Aber die *[deutet auf Typisch-Steckbrief]* sagen, dass es gleich, dass es ungefähr gleich ist.
79	K	That's a bit weird.	Irgendwie ist das n bisschen komisch.

Analysis of mathematizing pathway. In this phase of the design experiment session, the students first investigate the filled-in report sheets. Nesrin begins by noting that the Distance Report Sheet asserts that the ice "decreased" (#76), whereas the Typical Report Sheet asserts that it "roughly remains the same" (#76). From this, two theorems-in-action in her structuring activity can be reconstructed: *<MA/s: from 1982 to 2012, the general state of the ice decreased>* and simultaneously *<MA/s: from 1982 to 2012, the general state of the ice remains the same>*. Her reference to the Arctic sea ice seem to remain ambiguous, as Nesrin does not distinguish between different aspects of the phenomenon like the melting process and the winter months, but instead refer to an unspecified "it" (#76). From these theorems-in-action a concept-in-action can be derived which refers to a vague aspect of the ice, here referred to as ||*MCs/a: general state*||. Nesrin seems to perceive this aspect ||*MCs/a: general state*|| as represented by two different measures, Typical as well as the Distance. The conflicting assertions on the behavior of ||*MCs/a: general state*|| found in the report sheets, however, irritate the students. Kaan thus explains them by faulty calculations ("maybe they didn't measure correctly", #77): *<MA/f: either Typical or Distance were calculated wrong>*.

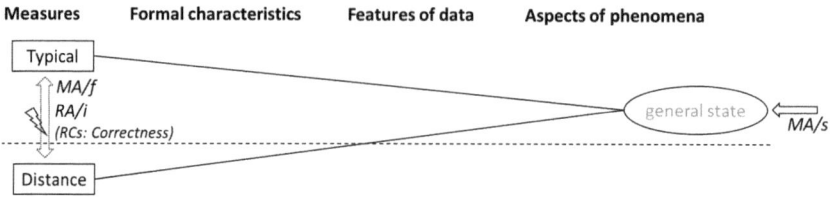

Fig. 8.3: Kaan and Nesrin represent the same aspect of the phenomenon with two different measures. The resulting conflict is interpreted as a conflict of formalizing. They also engage in the reflective activity of identifying patterns of thought

Analysis of reflective activities. In this scene, the students are in the reflective situation of rejecting measures. To Nesrin and Kaan in this moment, the conflicting summaries on the report sheets do not have their cause in different represented aspects of the phenomenon, but in the faulty calculation. They seem to follow the implicit assumption that, had the measures been calculated correctly, the summaries of the phenomenon would have been the same. To them, only an incorrect method of calculation can explain the differences in the summaries of the filled-in report sheets: "Maybe they didn't measure correctly" (#77). This shows an implicit identification of patterns of thought regarding the use of statistical measures: $<RA/i$: *Correct use of measures creates correct views on phenomena*$>$.

Analysis of reflective concepts. Behind their identifying activity, the situative reflective concept-in-action of $||RCs$: *correctness*$||$ can be reconstructed. They use this concept to argue that if one uses a measure in a 'correct' way (i.e. in accordance to its formal characteristics), one will arrive at the one 'correct' view of the phenomenon. They consider this as true also for other measures: used correctly, all measures provide the same correct view of the phenomenon. This situative concept resonates with a general reflective concept of $||RCg$: *consistency*$||$ of measures: mean, median, and mode generally produce mainly consistent results, as they are all measures of center. However, also inconsistencies between those measure can occur, mostly indicating special circumstances in the data distribution. In this part of the design experiment, the students have only identified a single (still rather unspecified) aspect of the phenomenon, represented by the different measures. Thus, for the students, it is highly astonishing that two measures would provide inconsistent results.

Arctic Sea Ice Problem – Episode 2

Some minutes later, the design experiment progresses to Phase 3, and the students are given the data of 2012. The design experiment leader reminds the students of their previous difficulties in interpreting the conflicting summaries and asks them whether they now find any explanation.

IV-2-KN; Phase 3; Start: 16:30
Previously, the students struggled to find an explanation for the conflicting filled-in report sheets. Now, given the data of 2012, the students identify several aspects of the phenomenon and use measures to represent them. Two conflicting aspects of phenomena, however, cause them to stop.

112	DL	Well, taking another look at these report, you said initially that it's a bit weird - with some saying it's changing a bit while others say it's rather not.	Ja und wenn man sich das jetzt nochmal anschaut, diese Steckbriefe anschaut, ihr habt am Anfang schon gesagt, das ist irgendwie komisch – die einen sagen, es ändert sich was und die anderen sagen, eher nicht.
113	K	#Yes.	#Ja
114	DL	And now we have the data. How, what does that mean now? What would you say as to why that is there?	Und jetzt haben wir die Daten. Wie, was heißt das denn jetzt? Warum ist das da, was würdet ihr dazu sagen?
115	N	Well, I think, with this here *[points to Typical Report Sheet]* they believe that in certain months - well - the volume of ice remains the same, but that it still sinks. Meaning, that it's the same for some months, only. Not even always.	Also ich glaube, bei dem hier *[deutet auf Typisch-Steckbrief]* meinen die, dass es in einigen Monaten – ja – an der Menge von Eis gleich bleibt, aber dann doch sinkt, also, dass es nur in einigen Monaten gleich ist. Und nicht immer.
116	K	Yes, and this one *[points at Distance Report Sheet]* nicely fits the statement that he, they say here as well that - the distance decreased by half - so the ice did sink by half. Which you can see here as well, where there were 10.5M km² but it's just 3.5M km².	Ja und der hier *[deutet auf Abstand-Steckbrief]* passt halt eben genau auf die Aussage, weil er, die sagen ja hier auch, dass – der Abstand ist um die Hälfte gesunken – Also das Eis ist um die Hälfte gesunken. Und hier kann man das eben auch erkennen, da waren 10,5 Mio. km² und hier sind es dann halt 3,5 Mio. km²
117	N	And that green one *[points to Value Report Sheet]* it's like - they say that it kept on retreating a bit in 2012, but it really retreated a lot. I don't quite understand why they picked just a little bit. Because it actually retreated by half., so it would be something like here *[points to middle of the sketch in Value Report Sheet]*	Und dieses Grüne *[deutet auf Wert-Steckbrief]* das ist ja – die sagen ja, dass es immer etwas zurückgegangen ist in 2012, aber eigentlich ist es doch sehr viel zurückgegangen. Ich versteh das dann nicht, warum die dann nur so ein bisschen genommen haben. Weil das ist ja eigentlich die Hälfte zurück gegangen, also müsste es dann ja irgendwie hier sein *[deutet auf Mitte der Skizze im Wert-Steckbrief]*

Analysis of mathematizing pathway. With access to the full data, the students begin by engaging in various mathematizing activities. Nesrin notes that the ice is "the same for some months, only". This is a much more distinguished view of the phenomenon than the students' previous concept-in-action of $||MCs/a:$ *general state*$||$. Thus, she engages in the mathematizing activity of structuring: $<MA/s:$ *Arctic sea ice comprises the ice being the same in some months>*. By relating to the Typical Report Sheet (#115) she also uses the situative measure of ||*MCs/m: Typical*|| to represent this aspect, engaging in the mathematizing activity of representing: <*MA/r: Typical represents the ice being the same in some months*> (#115).

Kaan follows in the structuring activity by identifying the aspect of the ice ||*MCs/a: retreating by half*|| ("the ice did sink by half", #116), although it is not exactly clear what he means by that. One interpretation would be that this aspect of the phenomenon corresponds to the minimum dropping from 10.5 to 3.5 million km² (#116, where Kaan seems to have read off a wrong value, as the minimum of 1992 is 7.5). By referring to the Distance Report Sheet, he engages in representing this aspect of the phenomenon through the situative measure ||*MCs/m: Distance*||.

Nesrin then examines the ||*MCs/m: Most Important Value*||. This again prompts a conflict. She finds that <*MA/r: the Most Important Value represents the ice retreating a bit*> ("it kept on retreating a bit in 2012", #117). Again, Nesrin does not further specify which feature of the data this aspect of the phenomenon corresponds to. This time, however, the conflict does not concern whether the ice melts at all, but in how much it melts: <*MA/s: Arctic sea ice cannot comprise both, the ice retreating a bit and the ice retreating by half*> ("I don't quite understand" the different summaries, #117). The students, however, do not seem to be able to resolve this conflict yet.

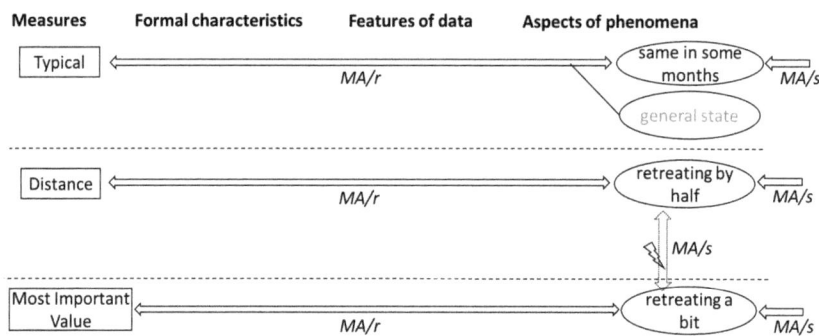

Fig. 8.4: Kaan and Nesrin engage in structuring by identifying three aspects of the phenomenon. They also engage in representing by linking the three situative measures with these three aspects. A conflict regarding structuring arises

Analysis of reflective activities. No reflective activity can be observed, as the students are not in a reflective situation.

Analysis of reflective concepts. No reflective concept can be identified.

Arctic Sea Ice Problem – Episode 3

In the following minutes (not shown here), the students start to favor the Distance Report Sheet, agreeing with the observed ||*MCs/a: retreating by half*||. However, they do not fully resolve the conflict between the Distance and Value Report Sheets. The design experiment leader follows up by challenging the students to evaluate the Value Report Sheet.

IV-2-KN; Phase 3; Start: 19:00
Previously, Kaan and Nesrin favored the Distance Report Sheet, but did not explain the conflicts between the filled-in report sheet. Now, they evaluate the Value Report Sheet and identify a new aspect of the phenomenon.

127	DL	So you'd say that this one is incomprehensible *[points Value Report Sheet]*. Or, would you say that it's actually wrong what they, what they claim?	Also würdet ihr sagen, man kann den nicht verstehen *[deutet auf Wert-Steckbrief]*, oder würdet ihr sagen, das stimmt gar nicht, was die da, was die da sagen?
128	N	and it's somehow wrong, because they say that 1982 is the mos-, they claim that the most important value is 16 and then 2012 has as most important value a 15.	Ich würde sagen, ich kann's nicht verstehen und irgendwie stimmt das auch nicht, weil die ja sagen, dass 1982 ist der wich-, die sagen ja, dass der wichtigste Wert 16 ist und dann 2012 ist der wichtigste Wert 15.
129	DL	Um.	Mhm
130	N	Oh, I believe that they mean to say that 15 is the most important value, that the ice moves up to 15, 15	Ach ich glaube, dass die damit meinen, der wichtigste Wert ist 15, dass das Eis bis zu 15 vordringt, 15
131	K	Km².	Km².
132	N	Right, km² - million Km² retreated again, so that it grew that much again. Like it says there. That's what I believe they meant. That's because, you can see that, that 15 *[hints at data from 2012 at approximately 15 million km²]*, well that this point, the yellow one, is at 15 and I believe that is then - the month where the, uh, ice retreats and that it retreats to 15. I believe that's what they mean here.	Ja km² - Mio km² wieder zurückgegangen ist, also dass es so groß wieder geworden ist. Wie es da steht. Ich glaub, dass die das so meinten. – Das ist da bei, das sieht man ja auch, dass 15 *[deutet im Diagramm von 2012 auf ca. 15 Mio. km²]* also dass der Punkt, der gelbe da, bei 15 liegt und ich glaube, das ist dann der – Monat, wo es äh das Eis wieder zurückgeht und dass es dann bis zur 15 wieder zurückgegangen ist. Ich glaub, das meinen die damit.
133	K	Yes.	Ja.

Analysis of mathematizing pathway. The conflict between the conflicting filled-in report sheets resurfaces as Nesrin seems to have troubles interpreting the ||*MCs/m: Most Important Value*|| (she "can't understand it", #128). Then however, she notes that there exists a point which the ice "increases up to" (#130), to which it goes "back to" (#132). This aspect of the phenomenon can be reconstructed as the ||*MCs/a: recovery point*|| of Arctic sea ice, the point in winter to where the ice recovers after melting in the summer. This shows her

mathematizing activity of structuring: instead of drawing on an unspecific
||*MCs/a: general state*|| of Arctic sea ice, she identifies this new aspect of the
phenomenon, presumably through her theorem-in-action <*MA/s: Arctic sea ice
comprises the recovery point*>. With the new aspect identified, she can immediately represent this aspect by noting that <*MA/r: the Most Important Value
represents the recovery point*> ("I think that is what they mean", #132), thus
ending the conflict between the two differing measures: the two measures represent different aspects, and thus the filled-in report sheets can come to different
summaries.

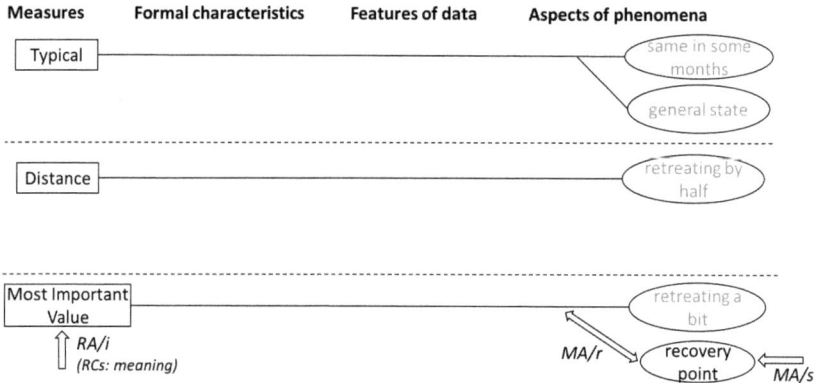

Fig. 8.5: Kaan and Nesrin identify a new aspect of the phenomenon and link it to the Most
Important Value, engaging in structuring and representing

Analysis of reflective activities. This Episode presents a reflective situation of
rejecting the measure Most Important Value. In this scene, Nesrin consciously
reflects on why the creators of the Value Report Sheet have such a different
view of the phenomenon than asserted by the Distance Report Sheet. The final
breakthrough is achieved by identifying what the creators "meant" (#130),
which leads her to the identification of a new aspect of the phenomenon. This is
not merely accidental discovery. Instead, this search for meaning constitutes a
very intentional activity pursued by Nesrin. This activity can be seen as the
reflective activity of identifying patterns of thought, because Nesrin actively
tries to find an intuitive meaning for the (to her) unintuitive measure of Most
Important Value. In her actions, she enacts the theorem-in-action <*RA/i: different measures can have a different meaning*>.

Analysis of reflective concepts. Nesrin's reflective activity helps her to differentiate between two different measures. Her situative reflective concept-in-
action of ||*RCs: meaning*|| allows her to consciously link measures to aspects of

phenomena. She is able to reflect on the fact that measures never represent (i.e. 'mean') whole phenomena, but only small specific aspects. Therefore, different measures can 'mean' different things, and for interpreting a measure, one has to identify the represented aspect of the phenomenon. Thus, $||RCs:\ meaning||$ could be a situative precursor to the general reflective concept of $||RCg:\ specificity||$: the idea that measures create partitions of phenomena and only represent specific aspects.

Arctic Sea Ice Problem – Episode 4

The identification of the new aspect proves to be a breakthrough for the students' reflective activities, as shown in the next sequence: Immediately following the last sequence, the students are asked to evaluate the report sheets.

IV-2-KN; Phase 3; Start: 20:10
Previously, Nesrin identifies the new aspect of the point where the ice goes back to. Now, she resolves the conflict between Distance and Value Report Sheets by differentiating between different aspects of the phenomenon to be addressed by the different measures.

134	DL	Um. Now is that - how is it when they - are they right, did they do this well, or don't you agree?	Mhm. Und ist das - wie ist das denn jetzt, wenn die - haben die jetzt recht, haben die das gut gemacht oder findet ihr das nicht so gut?
135	N	Well, it's okay. If you - you just need to think very carefully here *[hints at Value Report Sheet]*, and initially you don't get it right away, what they mean here. And, uh, this 15 is *[Most Important Value in 2012]* like - in the end - um - they could rather - uh - say what, what happens in the other months. I believe that would be more likely, So, if you	Also es geht eigentlich. Wenn man – bei dem muss man halt sehr viel nachdenken *[deutet auf Wert-Steckbrief]* und am Anfang, da versteht man das noch nicht sofort, was die damit meinen. Und ähm, diese 15 ist *[Wichtigster Wert in 2012]* halt – am Ende ja das heißt – ähm – die könnten auch – ähm – eher sagen, was in, was in den anderen Monaten passiert, ich glaub, das würde eher eintreffen. Also wenn man
136	DL	How do you mean?	Wie meinst du das?
137	N	Well, um, if you'd just - here *[Value Report Sheet]* you have 16 and there it's 15 *[Value Report Sheet]* and when you compare here *[hints at Distance Report sheet]*, they say that it retreated by half. And in particular here *[hints at Value Report Sheet]*] they say nothing about retreating or advancing, they just say that in the end the ice was this large.	Also ähm, wenn man halt – hier *[deutet Wert-Steckbrief]* hat man ja die 16 und da die 15 *[deutet Wert-Steckbrief]* und wenn man das aber in dem hier guckt *[deutet auf Abstand-Steckbrief]*, da sagen die, dass es um die Hälfte zurückgegangen ist. Und hier gar *[deutet auf Wert.-Steckbrief]* sagen die ja nichts mit zurückgehen und oder wieder nach vorne gehen, die sagen nur am Ende war das Eis wieder so groß.
138	DL	Mhm.	Mhm.
139	N	And they should have, I believe, simply added what happened in between.	Und die hätten, glaub ich, auch ruhig sagen können, was in der Mitte passiert ist.

Analysis of the mathematizing side. As a result of the previous structuring activity, instead of drawing on an unspecific ||*MCs/a: general state*|| in her structuring activity, Nesrin draws on two different concepts-in-action to structure the phenomenon into two distinct aspects of the phenomenon: the ||*MCs/a: recovery point*|| (where the ice is "in the end", #137) and the ||*MCs/a: melting process*|| (the "retreating" and "advancing" of ice within one year, #137, #139). Her contrasting of the Distance Report Sheet with the Value Report Sheet while identifying the aspects of the phenomenon (#137) can be interpreted as a mathematizing activity of representing: <*MA/r: the Distance represents the melting process*> and <*MA/r: the Most Important Value represents the recovery point*>.

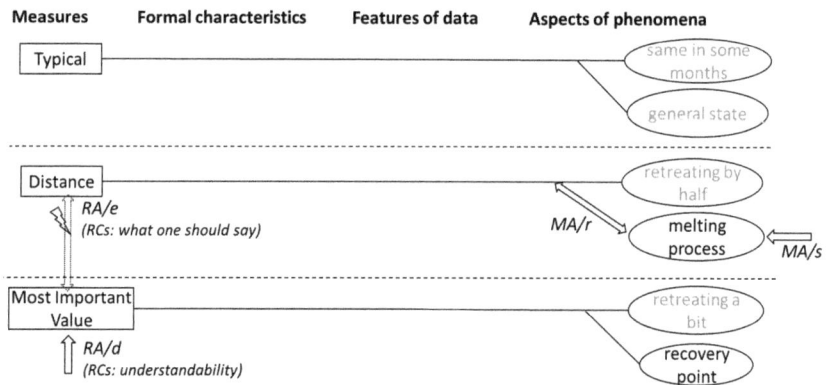

Fig. 8.6: Nesrin identifies a new aspect of the phenomenon and represents it through the Distance. She also engages in reflective activities, criticizing the Most Important Value by denominating risks and limits and contrasting two situative measures through their aims and purposes

Analysis of reflective activities. Nesrin begins to contrast the different measures, showing many signs of reflective activities. In the beginning, she gives clear precedence to the Distance Report Sheet over the Value Report Sheet, because with the Value Report Sheet "you just need to think very carefully here, and initially you don't get it right away, what they mean here" (#135). In this way, Nesrin does not refer to any difference concerning the mathematizing activities, but instead engages in reflective activity. She denominates a limit of the situative measure of the ||*MCs/m: Most Important Value*||: <*RA/d: the Most Important Value is not easily understandable*>.

She also gives clear preference to the ||*MCs/m: Distance*|| over the ||*MCs/m: Most Important Value*||. To her, the different measures do not simply just represent different aspects of the phenomenon. Instead, one has to choose a measure that provides evidence for the intended argumentation; the measure has to fit one's aims. Thus, she engages in the reflective activity of explicating aims and purposes: *<RA/e: the Distance should be used instead of the Most Important Value for arguing about Arctic sea ice>* (they should rather have said what happened in the middle, #139).

Analysis of reflective concepts. Nesrin criticizes the Most Important Value in terms of ||*RCs: understandability*||: it is not easily understandable, whereas the Distance is (#135). The point of measures, however, is to communicate understandings of phenomena. Therefore, the Distance is the better measure. Thus, ||*RCs: understandability*|| could be a situative precursor for the concept of ||*RCg: intersubjectivity*||: general measures enable intersubjective communication about phenomena, because their formal characteristics ensure a common ground for all interlocutors.

In the previous scene, Nesrin also already drew on a situative precursor to the ||*RCg: specificity*|| of measures. Now, she is able to give preference to some of those specific aspects. She compares the two measures and their represented aspects after ||*RCs: what one should say*||, possibly a situative starting point for ||*RCg: contextual relevance*||: not only do measures only address specific aspects, but one has to decide which aspect actually should be addressed.

Arctic Sea Ice Problem – Episode 5

The design experiment progresses through Phase 4, and the students create their own report sheet (Fig. 8.7). Afterwards, when they are asked to explain and justify their report sheet, Nesrin recounts her process of comprehending the filled-in report sheets and emphasizes the impact of the reading order for her evaluation of the report sheets.

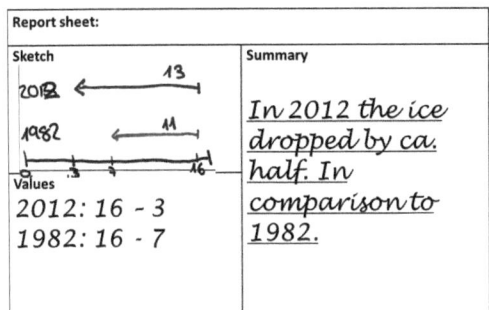

Fig. 8.7: Kaan and Nesrin's own report sheet

Previously, the students created their own report sheet. Now, Nesrin evaluates the filled-in report sheets by hypothesizing how her reading order of the filled-in report sheets has influenced her understanding of the phenomenon.

210	N	I think - somehow - that wasn't crystal clear *[points at Typical Report Sheet]* - as far as we're concerned - um, and it wasn't quite the correct description. The dots were accurately placed but it looked wrong, because when you - well, if you had read this first *[points at Distance Report Sheet]* it's like *[points to Typical Report Sheet]*], you don't believe it's right anymore.	Ich glaub – irgendwie – das ist nicht so einleuchtend *[deutet auf Typisch-Steckbrief]* gewesen – also für uns jetzt – ähm und das war auch nicht so die ganz richtige Beschreibung. Die Punkte waren zwar richtig gesetzt aber das sah irgendwie falsch aus, weil wenn man – also wenn man vorher das gelesen hat *[deutet auf Abstand-Steckbrief]* ist das irgendwie *[deutet Typisch-Steckbrief]*, findet man das nicht mehr so richtig
211	DL	And if you'd change the order of reading?	Und wenn man es andersrum liest?
212	N	Then it's probably a bit different, because if you'd here *[hints at Typical Report Sheet]*, well, I did read that one first *[hints at Distance Report Sheet]* and then this one *[hints at Typical Report Sheet]* and then I thought to myself, uh, that doesn't work somehow when they say that it is by half *[hints at Distance Report Sheet]* and first this *[hints at Typical Report Sheet]* and then that *[hints at Distance Report Sheet]*, then I don't think it would be like this if you had read that first *[hints at Distance Report Sheet]*.	Dann ist das wahrscheinlich ein bisschen anders, weil wenn man da *[deutet auf Typisch-Steckbrief]* also ich hab ja das zuerst gelesen *[deutet auf Abstand-Steckbrief]* und dann *[deutet auf Typisch-Steckbrief]* das und dann dacht ich mir nur so hä irgendwie geht das doch gar nicht, wenn die sagen, das ist erst zur Hälfte *[deutet auf Abstand-Steckbrief]* und wenn man erst das *[deutet auf Typisch-Steckbrief]* und dann das *[deutet auf Abstand-Steckbrief]*, dann wär das glaub ich nicht so gewesen, wie wenn man erst das gelesen hätte *[deutet auf Abstand-Steckbrief]*.
213	DL	Um, why? What would you have likely thought then?	Mhm wieso, was hättest du denn dann gedacht eher?
214	N	If I had read that only now, well *[hints at Typical Report Sheet]*, that it remains about the same, it remains about the same, you can actually see that *[points to diagram]*, it's all good. Then I take a look at this *[hints at Value Report Sheet]* and then that *[hints at Distance Report Sheet]*, with it decreasing by 50%, then somehow it's already in my head what I read first, somehow. And somehow after reading it twice, three times again, I'd get that idea that, um, it's rather a bit weird reading that *[hints at Distance Report Sheet]* but I'd get it only a little later.	Wenn ich jetzt erst das gelesen hätte, also *[deutet auf Typisch-Steckbrief]* dass es so ungefähr bleibt, das bleibt ungefähr, ja sieht man ja auch *[deutet auf Diagramm]*, alles okay. Dann guck ich mir das an *[deutet auf Wert-Steckbrief]* und dann das *[deutet Abstand-Steckbrief]*, dass es um 50% weniger wird, dann irgendwie, in meinem Kopf ist dann schon drin, was ich als erstes gelesen hab irgendwie. Und irgendwie nach zwei, dreimal nochmal lesen, würde ich dann erst auf die Idee kommen, dass ähm dass es eben ein bisschen komisch ist, wenn man das da liest *[deutet auf Abstand-Steckbrief]* aber erst ein bisschen später würd ich glaub ich drauf kommen.

Analysis of mathematizing pathway. It is worth noting that in this scene, although Nesrin shows some rich reasoning, this does not concern the mathematizing side of mathematical literacy. She does not identify any new aspects of the phenomenon, features of the data, or formal characteristics, although she is investigating the situative measure $||MCs/m: Typical||$. It seems that her concepts-in-action $||MCs/a: same in some moths||$ represented by $||MCs/m: Typical||$ and, $||MCs/a: retreating by half||$ represented by $||MCs/m: Distance||$ suffice. Instead of engaging in mathematizing activities, she is fully committed to reflecting on statistical measures.

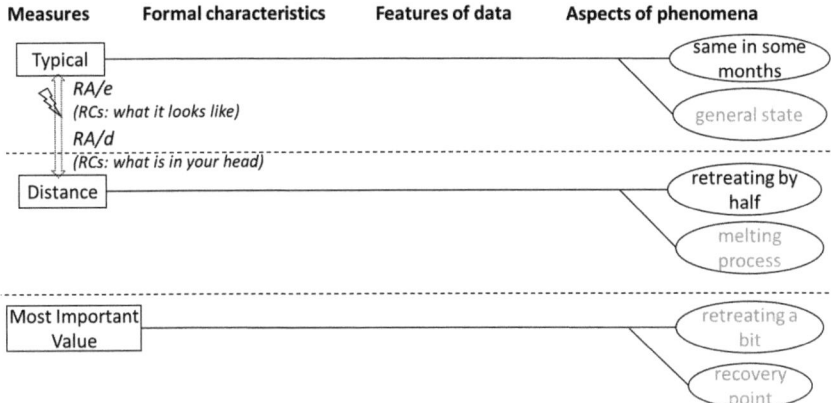

Fig. 8.8: Nesrin engages in reflective activities to contrast the two situative measures of Typical and Distance

Analysis of reflective activities. On the reflective side however, Nesrin shows rich reflective activities. In this scene, she describes the impact that her reading order of the filled-in report sheets had on her understanding of the phenomenon: if she had hypothetically read the Typical Report Sheet first, she would have accepted its summary after a brief look at the data (simulating checking the data: "it remains about the same, you can actually see that", #214). Only after her views had already been formed, she would have read the Distance Report Sheet. When finally reading the Distance Report Sheet, she would already have been so convinced by the Typical Report Sheet that she would have needed to read the Distance Report Sheet several times to comprehend, and ultimately prefer, the view of the phenomenon presented by the Distance Report Sheet ("but I'd get it only a little later", #214).

Nesrin's actions can be interpreted as engaging in extensive reflective activities. Her reflection of the reading order is based on the fact that the different measures can be used to make different aspects of the phenomenon visible. Her theorems-in-action explicate the purposes of the measures that <*RA/e: using Typical, the Arctic sea ice looks like it is staying the same*>, whereas <*RA/e: using the Distance, the Arctic sea ice looks like it is declining*>. This, however, is not a simple difference in perspective, but shows a major risk: one should use the Distance, but using Typical, one would arrive at plausible but misleading inferences. By referring to her reading order, she illustrates what is "in her head" (#214) when reading the filled-in report sheets, which she frames as a clear risk. Thus, her theorem-in-action concerns the denomination of the risks of using measures for investigating phenomena: <*RA/d: if Typical is in your head, you do not accept the Distance*>, and this can result in discarding the Distance Report Sheet, although it would ultimately be preferable for Arctic sea ice.

Analysis of reflective concepts. Earlier, Nesrin reflected on measures in terms of ||*RCg: consistency*||, under which inconsistent perspectives of measures hint at some fundamental problem like faulty calculation. The reflective concept of ||*RCg: consistency*|| alone, however, cannot be applicable here (as the measures were used correctly, #210). In this scene, Nesrin seems to overcome her reflective concept of ||*RCg: consistency*|| in favor of differentiating in terms of ||*RCs: what it looks like*|| when using a certain measure (#214). In this way, she draws on a situative precursor to the general reflective concept of ||*RCg: perspectivity*||: both measures might be used correctly, they just provide different perspectives on the phenomenon.

This again is not just taken as given, but judged critically. By tracking the impact of what she read first, Nesrin illustrates how her own perceptions can be formed by ||*RCs: what is in your head*||: although ||*MCs/m: Typical*|| ultimately is not preferable, it was so convincing to change her perception of the other report sheets. In this way, she even identifies the ||*RCg: formatting power*|| of mathematics, as Typical creates a single perspective that seems objectively true where there is, in fact, a plurality of perspectives through other measures.

Summary for Kaan and Nesrin

Throughout the design experiment, Kaan and Nesrin draw on a variety of situative reflective concepts. Their discussion starts with the conflict between summaries provided by the different filled-in report sheets (Fig. 8.9). In the beginning, they explain those differences through errors in calculation due to their concept-in-action of ||*RCs: correctness*||, as they only identified a single unspecific aspect of ||*MCs/a: general state*||. Later, they manage to structure the phenomenon by differentiating between the aspects of ||*MCs/a: recovery point*|| and ||*MCs/a: melting process*|. This enables them to draw on ||*RCs: what it looks like*|| in order to relate those differences to differences in ||*RCg: perspectivity*||. In the end, their concept-in-action of ||*RCs: what is in your head*|| even gives first ideas of the general concept ||*RCg: formatting power*|| that can be projected through measures.

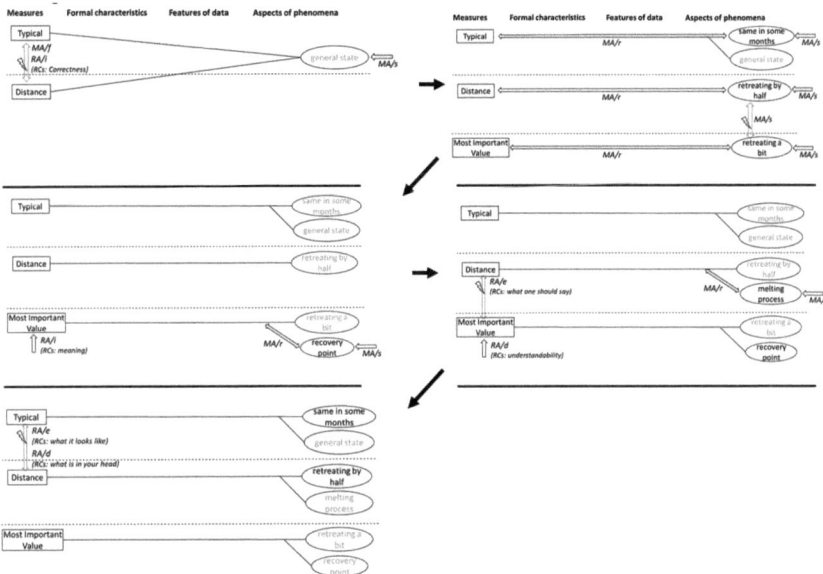

Fig. 8.9: Overview on Kaan and Nesrin's learning process

While they engage in various reflective activities, their mathematizing activities show a focus on structuring and representing (Fig. 8.10). Again, the students' mathematizing activities coincide with their rich reflective activities. Their network of reflective concepts is systematized in Section 8.3.

Mathematizing activities			Reflective activities		
Structuring	Representing	Formalizing	Identifying	Making explicit	Denominating
			Arctic Sea Ice Task		
			Correctness		
			Meaning		
				What you should say	Understandability
				What it looks like	What is in your head

Fig. 8.10: Mathematizing and reflective activities of Kaan and Nesrin during the Arctic Sea Ice Problem

As already observed in the design experiments with Maria and Natalie (Section 7.2.1), the mathematizing activity of structuring seems to be closely linked to reflective activities. Additionally, Kaan and Nesrin often simultaneously engaged in structuring and representing. It seems that both mathematizing activities can provide a cause for various reflective activities. Compared to the rich reflections of Maria and Natalie, however, Nesrin shows an even wider network of different reflective concepts in her reflective activities. She thus seems able to purposefully draw on nuanced reflective concepts depending not on the single activity at hand, but on the reason for her reflection.

8.2.2 Jana and Mara: Judging the contextual relevance of measures

Report sheet: *Troll Forskningsstasjon*	
Sketch	**Summary**
very frequent ![sketch]	*Sometimes it is -23°C cold but most of the time it*
Values *frequent: -19 - -9°C*	*is -19 - -9°C cold. But there also are exceptions from -6 - -5°C*

Fig. 8.11: Jana and Mara's report sheet from the Antarctic Temperatures Problem

The second case shows the seventh graders Jana (J) and Mara (R). Both students are highly motivated, although Jana seems a bit shy. During the design experiment session, they begin to interpret the word 'important' in Most Important Value, leading to reflections on contextual relevance.

The analysis begins in the middle of the second design experiment, in Phase 3, in which the students evaluate the filled-in report sheets based on of data from 1982, 1992, and 2012. Until now, the students did not yet choose a report sheet they fully agree with. The analysis starts as the design experiment leader asks them to evaluate the Value Report Sheet.

IV-2-JM; Phase 3; Start: 24:10
Previously, the students acknowledged the different filled-in report sheets, but did not focus on the conflicting summaries. Now, with the data of 2012, Jana and Mara explain why they do not agree with the Most Important Value.

139	DL	What exactly is this one about *[hints at Value Report Sheet]*? You just didn't agree to the most important values.	Was ist denn jetzt genau mit dem hier? *[deutet auf Wert-Steckbrief]* Gerade warst du da mit den wichtigsten Werten nicht einverstanden.
140	J	Well, because it's actually the wrong term. The highest value, that would, actually ...	Ja, weil das eigentlich der falsche Begriff ist. Der höchste Wert, das würde, eigentlich...
141	M	Well, the most important value - it really is about how much it retre... well, how much it shrunk. And not	Ja der wichtigste Wert – ist ja wirklich jetzt wie stark das zurückge– also, wie stark das geschrumpft ist.

| how high the ice was, well - it's still important to know how high the ice was, but - well, how much ice there was, um, but - I think, like, that this is not the most important value. The most important value would be, I believe, the lowest value - no - that one isn't most important either. Well, I'd have written down lowest and highest values over the years. That would have been a lot simpler. | Und nicht wie hoch das Eis war, also – es ist schon auch wichtig, zu wissen, wie hoch das Eis war, aber – also wie viel Eis da war, - ähm aber – ich find es ist glaub ich nicht der wichtigste Wert. Der wichtigste Wert wäre, glaub ich, der niedrigste Wert – nein – der ist auch nicht am wichtigsten. Also ich hätte niedrigster und höchster Wert aufgeschrieben. in den Jahren. Das wär deutlich einfacher. |

Analysis of mathematizing pathway. The students now make their view of the phenomenon explicit. When speaking of Arctic sea ice, Mara distinguishes between "how much it shrank" and "how high the ice was" (#141). These can be interpreted as two aspects of the phenomenon: the $||MCs/a: high point||$ of the winter months to where the ice recovers, and the $||MCs/a: ice melt||$ within one year, the departure of the $||MCs/a: high point||$. Mara draws on these concepts-in-action in her theorem-in-action $<MA/s: Arctic sea ice comprises the high point and the ice melt>$ (#141), showing her structuring activity. Whereas the former aspect is represented by the $||MCs/m: Most Important Value||$, the latter (the "really" most important, #141) starts out missing a representing measure. Here, Mara engages in representing: The aspect $||MCs/a: ice melt||$ is not represented by the lowest value alone (", the lowest value - no - that one isn't most important either" while searching for actually important values, #141). Instead, comparing the highest and lowest values does: $<MA/r: the highest and lowest values represent the ice melt>$ To Mara, this seems to be different to the measure $||MCs/m: Distance||$, since she does not refer to the formal characteristic of taking the difference between those values.

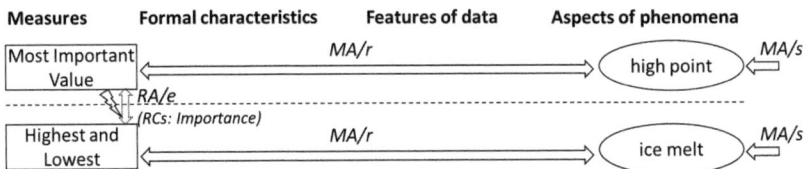

Fig. 8.12: Jana and Mara engage in structuring by identifying two aspects of the phenomenon and represent them through two different measures. They engage in explicating aims and purposes to contrast the two measures

Analysis of reflective activities. Jana and Mara take the term 'Most Important Value' literally, objecting to the claim that the situative measure of Most Important Value actually is important ("it's actually the wrong term", #140 and "this is not the most important value", #141). Mara instead proposes their own measure of the $||MCs/m:$ *Highest and Lowest Values*$||$. For her justification, she explicitly takes the phenomenon of Arctic sea ice into account, arguing why other measures do not fit the purpose at hand. Her actions can be interpreted as engaging in the reflective activity of explicating aims and purposes: <*RA/e: for arguing about Arctic sea ice, the Highest and Lowest Values are more important than the Most Important Value*>, because they address the aspect of the $||MCs/a:$ *ice melt*$||$, which is more important than the $||MCs/a:$ *high point*$||$.

Analysis of reflective concepts. In her theorem-in-action regarding which measure should actually be used for Arctic sea ice, Mara draws on the situative reflective concept-in-action of $||RCs:$ *importance*$||$. One could possibly choose any measure and use it to investigate Arctic sea ice. However, one should actually use one that is important. According to Mara, it is important to represent the $||MCs/a:$ *ice melt*$||$ (#141). This situative concept of $||RCs:$ *importance*$||$ can serve as a situative precursor to the general concept of $||RCg:$ *contextual relevance*$||$. For Arctic sea ice, it is simply more important to address the melting process of ice, identified here as $||MCs/a:$ *ice melt*$||$. In general, measures differ in their contextual importance depending on the phenomenon under investigation.

Arctic Sea Ice Problem – Episode 2

Although the students differentiate between aspects, they do not address and explain the different summaries of the filled-in report sheets. A short bit later, the design experiment leader challenges them to explain these differences.

IV-2-JM; Phase 3; Start: 26:15
Previously, the students identify different aspects of the phenomenon, but do not explain the different summaries on the filled-in report sheets. Now, Mara engages in structuring and representing for investigating the use of the measures Typical and Distance.

157	DL	Okay. Nevertheless, that one *[points to Value Report Sheet]* claims that not much has changed and this one *[points to Distance Report Sheet]* claims that there were quite some changes. Which one is right?	Okay. Aber trotzdem sagt der *[deutet auf Wert-Steckbrief]* so viel hat sich da nicht geändert und der *[deutet auf Abstand-Steckbrief]* sagt, da hat sich schon was geändert. Was ist es denn jetzt?
158	M	Uh, something did change.	Ähm, es hat sich schon was verändert.
159	J	Um.	Mhm.
160	DL	Um.	Mhm.
161	J	In particular about distance *[points to diagram of 2012]*.	Vor allem an dem Abstand *[deutet auf das Diagramm von 2012]*.
162	M	Yes. So, moving from 1982 *[points to diagram]* to 2012 *[points to diagram]* that is quite something - as you can see here *[points to worksheet pictures]*. That is - that half there - well, it like totally	Ja. Also wenn man jetzt von 1982 *[deutet auf das Diagramm]* zu 2012 *[deutet auf das Diagramm]* ist das ja schon einiges – hier sieht man das ja auch *[deutet auf die Bilder auf dem Arbeitsblatt]* das ist - die Hälfte da –

	retrea.. uh, melted. That doesn't mean, uh - it remained about the same.	also ja also es ist total viel zurück– äh geschmolzen, das heißt nicht, äh – das ist ungefähr gleich geblieben.
163	DL	Um.	Mhm
164	M	Well, the area with the most ice remained the same, that's true. That is, probably supposed to be the typical range - uh, so if that one *[points to Typical Report Sheet]* refers to typical range in an area with - uh, the most - ice, that means that it's correct, which is why the statement is correct too, right? That's really weirdly written.	Also der Bereich, wo am meisten Eis ist, ist gleich geblieben, das stimmt. Das ist, soll wahrscheinlich der typische Bereich sein – ähm, also wenn er *[deutet Typisch-Steckbrief]* mit typischen Bereich den Bereich, wo – ähm am meisten – Eis ist, meint, dann ist das richtig, aber dann stimmt die Aussage auch, oder? Das ist irgendwie auch komisch aufgeschrieben.

Analysis of mathematizing pathway. The students now turn towards the $||MCs/m: Distance||$ and $||MCs/m: Typical||$. In the beginning, Mara seems to have some difficulties in articulating how the two filled-in report sheets fit together ("much did reced- ehm melt, that doesn't mean, ehm – it roughly stayed the same", #162). Thus, she seems to identify the aspect of the ice $||MCs/a: retreating by half||$ ("that half there - well, it like totally retrea.. uh, melted", #162), represented by the $||MCs/m: Distance||$, which stands in conflict with the aspect of the ice $||MCs/a: staying the same||$ ("the area with the most ice remained the same, that's true", #164). She finds one explanation by the activity of representing, that $<MA/r: Typical represents the ice staying the same>$ (#164). However, she does not seem to be satisfied with that explanation ("That's really weirdly written, #164).

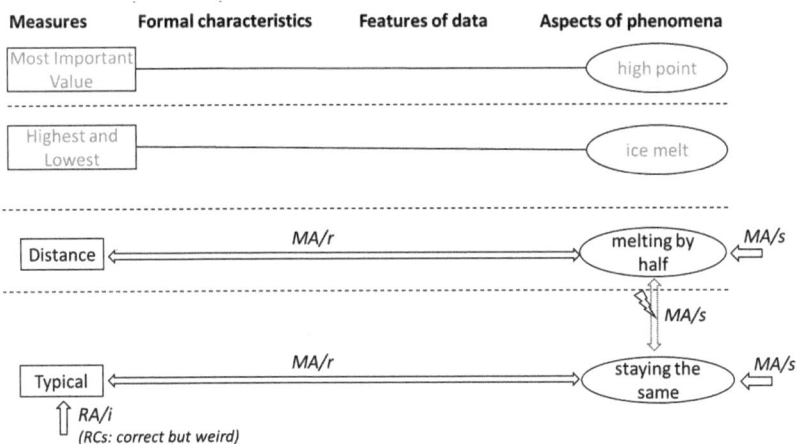

Fig. 8.13: Jana and Mara identify another two aspects of the phenomenon, which stand in conflict with each other. Mara engages in reflective activity concerning the situative measure Typical

Analysis of reflective activities. While comparing the filled-in report sheets, Mara notes that the use of ||*MCs/m: Typical*|| to represent the ice ||*MCs/a: staying the same*|| might be correct ("the statement is correct too", 164), but still shows signs of discomfort. This way of investigating the use of Typical does not rely on mathematizing activities, as to her, everything is "correct". Thus, her assessment that Typical might be technically correct, yet still somehow misleading, can be interpreted as a reflective activity of identifying patterns of thought: <*RA/i: Typical is correct, but weird*> (#164).

Analysis of reflective concepts. In her reflective theorem-in-action concerning patterns of thought for ||*MCs/m: Typical*||, Mara draws on her situative reflective concept-in-action of ||*RCs: correct but weird*||. Mara seems to acknowledge that the use of ||*MCs/m: Typical*|| and the assertion of the ice ||*MCs/a: staying the same*|| are technically correct, yet to Mara this still seems 'weird'. This could be a precursor to the general reflective concept of ||*RCg: perspectivity*||: using ||*MCs/m: Typical*|| could be correct, yet of little ||*RCs: importance*||. It creates a view of the phenomenon that, although correct, seems strange. Thus, ||*MCs/m: Typical*|| needs to be criticized on the level of ||*RCg: perspectivity*||

Arctic Sea Ice Problem – Episode 3

The design experiment progresses, and the students create their own report sheet which heavily draws on ideas presented in the Distance Report Sheet (Fig. 8.14). Afterwards, the students are asked to justify their choices made.

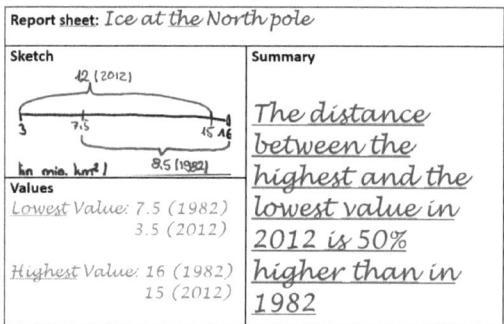

Fig. 8.14: Jana and Mara's report sheet

IV-2-JM; Phase 4; Start: 40:20
Previously, Jana and Mara created their own report sheet. Now, they explain their own report sheet and why they did not use the situative measure Typical.

#		English	German
230	DL	Okay, outstanding. You really worked hard here. Awesome. You just need to explain this to me. Why are doing it like this?	Okay, super. Da habt ihr jetzt noch richtig, richtig viel dran gearbeitet. Schön. Das müsst ihr mir jetzt aber noch erklären. Warum macht ihr das denn jetzt so?
231	M	Um.	Ähm
232	J	Because as accurate as possible - we wanted to use the sketch to accurately explain the distance.	Weil wir möglichst den – also bei der Skizze haben wir den möglichst den Abstand genau erklären wollten.
233	M	And because the distance is larger then there is less ice as well.	Und weil der Abstand größer ist, ist ja auch weniger Eis da.
234	J	Yes, exactly.	Ja genau.
235	M	So, then overall - less ice #and	Ähm also insgesamt dann – weniger Eis #und
236	DL	#Um.	#Mhm.
237	M	And that is somewhat similar *[points to Distance Report Sheet]*, but it was totally difficult to detect, which is why we have this *[points to own report sheet]*, so that, then, that you can tell better from this and make a comparison - directly, it's bit like - you don't really notice the other numbers *[points to Distance Report Sheet]*, but here you do *[points to own report sheet]*.	Und das ist ja schon so ähnlich *[deutet auf Abstand-Steckbrief]*, aber das konnte man so total schlecht erkennen, deswegen haben wir das so *[deutet auf eigenen Steckbrief]*, dass das dann das man das so besser erkennen kann und das dann auch im Vergleich sehen kann – direkt, da ist das so ein bisschen - da sieht man die anderen Zahlen gar nicht *[deutet auf Abstand-Steckbrief]* so richtig und dafür da *[deutet auf eigenen Steckbrief]*.
238	DL	Okay.	Okay.
239	M	And – this here *[hints at Value Report Sheet]* we thought that a most important value was rather daft and left it at lowest and highest value, there as well, uh, for 1982 and 2012, always. Yes.	Und – hier *[deutet auf Wert-Steckbrief]* fanden wir das ja doot mit dem wichtigsten Wert und haben das bei niedrigster und höchster Wert da auch noch äh immer für 1982 und 2012 – Ja.
240	DL	Um, I did understand that. Last time you gave some range. You didn't call it typical but something I don't remember exactly. When you were talking about temperatures. I think you identified some range, like where there were lots of dots. You omitted that just now. Why wasn't that important to you here?	Mhm, das hab ich verstanden. Letztes Mal habt ihr auch so einen Bereich gegeben, ihr habt das nicht typisch genannt, sondern ich weiß nicht mehr wie ihr das genau genannt hattet. Als ihr so über Temperaturen gesprochen hattet. Ich meine ihr hättet da so einen Bereich angegeben, so wo viele Punkte waren. Das habt ihr jetzt weggelassen. Warum war das denn hier für euch nicht so wichtig?
241	M	Um - because - because with temperature we were supposed to identify for which temperatures they should prepare most - and here, uh, it's a bit more important to see how the ice melted and not where the most - well, when, in what time - well - um - how much ice is there - the most.	Ähm – weil – weil wir bei der Temperatur ja gucken sollten für welche Temperaturen sie sich am meisten rüsten sollen – und ähm hier ist es, hier ist ja eher wichtig zu sehen, wie das Eis geschmolzen ist und nicht wo am meisten – also wann, in welcher Zeit – also – ähm – wie viel Eis am meisten – da ist.

Analysis of mathematizing pathways. In their report sheet, Jana and Mara use their situative measure ||*MCs/m: Highest and Lowest Values*|| to represent the aspect ||*MCs/a: ice melt*||. They then use their measure to represent changes in the phenomenon of Arctic sea ice, namely <*MA/r: the increasing distance between the Highest and Lowest Values represents the increasing ice melt*>. When asked why the students did not use ||*MCs/m: Typical*||, Mara refers to the phenomenon of Antarctic temperatures (the "temperatures", #241). In the previous session, the students used the measure ||*MCs/m: Typical*|| to represent a range of ||*MCs/a: expected temperatures*||. Trying to construct a similarity between the two phenomena, Mara meets difficulties in expressing what an aspect like a 'typical area of ice' could be (stuttering with "when, in what time - well - um - how much ice is there - the most", #241). Thus, she engages in structuring, identifying the aspect of the ||*MCs/a: period of most ice*|| (#241), represented by ||*MCs/m: Typical*||. This aspect of the phenomenon, however, is not important to her, so that the students chose ||*MCs/m: Lowest and Highest Values*|| as measure on their filled-in report sheet.

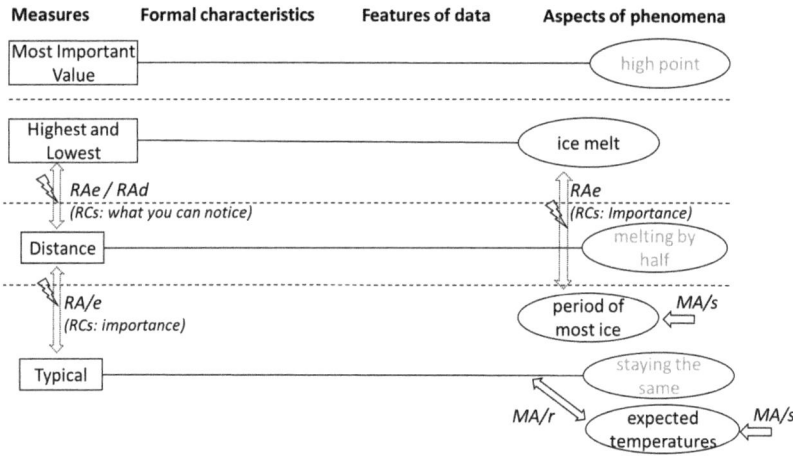

Fig. 8.15: Jana and Mara identify two new aspects of phenomena represented by Typical, one of them belonging to the phenomenon of Antarctic temperatures. They contrast situative measures through various reflective activities.

Analysis of reflective activities. Again, this Episode presents an excerpt where most of the activities fall on the reflective side of mathematical literacy. Mara starts by contrasting their measure of ||*MCs/m: Highest and Lowest Values*|| with the ||*MCs/m: Distance*||. With the aspects of ||*MCs/a: ice melt*|| and ||*MCs/a: retreating by half*||, both measures represent aspects concerning the melting process or Arctic sea ice. Yet she prefers their own measure, since it better suits her aims and purposes: <*RA/e: the Highest and Lowest Values can be used to better tell about how much the ice shrank*> ("you can tell better from this and make a comparison - directly", #237). The $||MCs/m: Distance||$ on the other hand even presents a risk, as the Distance Report Sheet conceals some information so that <*RA/d: the Distance cannot be used to easily notice how much the ice shrank*> ("you don't really notice the other numbers", #237). For justifying the choice of their own measure over $||MCs/m: Typical||$, Mara contrasts the different phenomena of Arctic sea ice and Antarctic temperatures, drawing on her theorem-in-action <*RA/e: for Antarctic temperatures, Typical is more important than the Lowest and Highest Values*>, whereas <*RA/e: for Arctic sea ice, the Lowest and Highest Values are more important than Typical*>.

Analysis of reflective concepts. As before, Mara draws on the situative reflective concept-in-action of $||RCs: importance||$ when explicating the aims and purposes of Typical. In this, her situative reflective concept-in-action shows some stable properties, as it also influenced the previous scenes of the learning process. Additionally, she contrasts the Lowest and Highest Values and Distance not by especially important represented aspects of the phenomenon, but by considering $||RCs: what you can notice||$ about the Arctic sea ice when utilizing a particular measure. This seems to be a ways in which she contrasts the two measures in terms of $||RCg: perspectivity||$. This comparison also presents a special way. Previous examples concerning the perspectivity of measures were related to different represented aspects of the phenomena. For Mara, the difference is found in their effectiveness: the $||MCs/m: Lowest and Highest Values||$ are more suited to provide $||RCg: Insight||$ into phenomena, as they generate more knowledge about Arctic sea ice than the $||MCs/m: Distance||$.

Summary for Jana and Mara

Whereas the students begin by focusing on the Most Important Value, the students' focus quickly changes to other measures (Fig. 8.16). As with Kaan and Nesrin, Jana and Mara also engage in contrasting different measures, often coinciding with structuring activities.

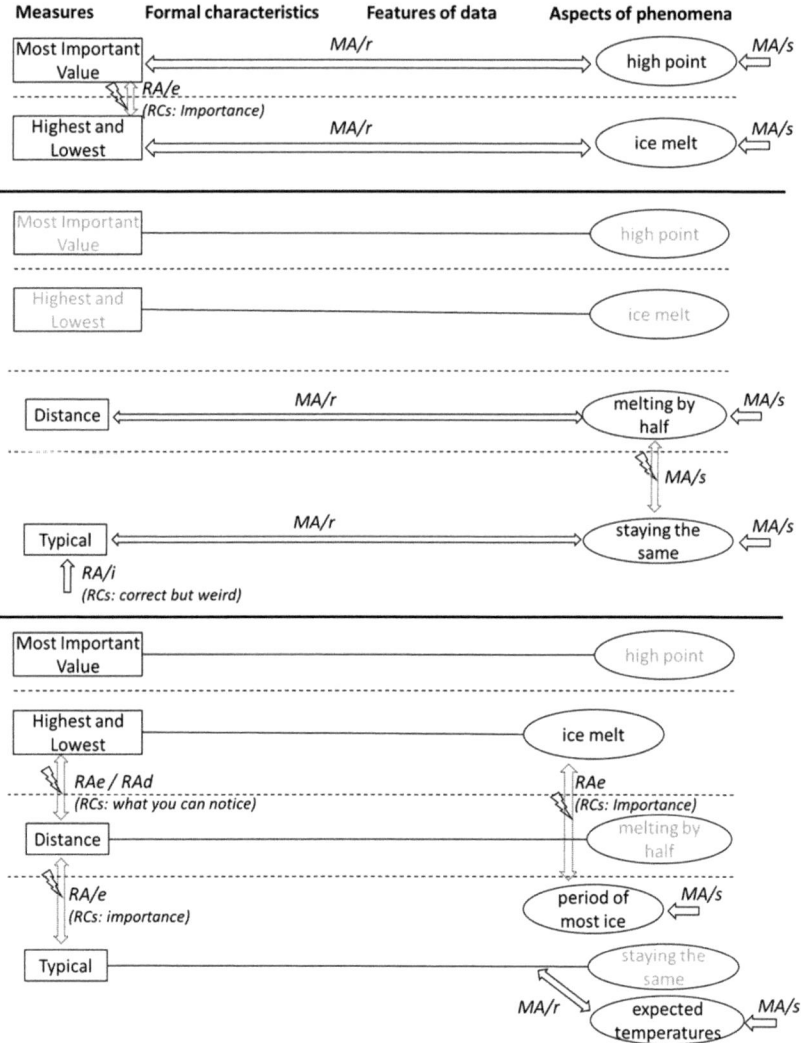

Fig. 8.16: Mathematizing and reflective activities of Jana and Mara

Jana and Mara also combine mathematizing and reflective activities. This pair stands out for their use of the situative reflective concept of $||RCs: Importance||$ (Fig. 8.17). Not only do the students draw on this concept to justify their choice of measure, it also helps them to compare different phenomena: what is important for Antarctic temperatures need not be important for Arctic sea ice, and thus one should use different measures. Through their use of $||RCs:$ *importance*$||$, the students are able to draw parallels or to explicate differences between phenomena. The use of specific measures for specific contexts gets more pronounced. In this way, they broaden the contextual neighborhood of $||MCs/m:$ *Typical*$||$ by drawing similarities and identifying differences between phenomena, and finding conditions under which $||MCs/m:$ *Typical*$||$ can be used in different contexts.

Mathematizing activities			Reflective activities		
Structuring	Representing	Formalizing	Identifying	Making explicit	Denominating
Arctic sea ice task					
				importance	
			correct but weird		
				importance	what you can notice

Fig. 8.17: Mathematizing and reflective activities of Jana and Mara

As with Kaan and Nesrin, Fig. 8.17 again shows a high coincidence of the mathematizing activities of structuring and representing with the different reflective activities. One reason for this could be the prominent reflective concept of $||RCs:$ *importance*$||$. As a situative precursor to $||RCg:$ *contextual relevance*$||$, it seems reasonable that such a concept might support the mathematizing activity of structuring. The $||RCs:$ *importance*$||$ of specific aspects of the phenomenon plays a large part in determining how to structure the phenomenon and which measure to use to represent these aspects of the phenomenon.

8.2.3 The effects of design for Cycle IV

The design changes implemented for Cycle IV aimed at creating more opportunities in which students could engage in reflective activities. Cycle III already showed how contrasting measures can elicit reflective activities. This resulted in the revised design principle DP/Measures-Revised, which placed a strong emphasis on eliciting reflective activities by contrasting different measures. This design principle also was introduced in order to reduce the role of the situative measure Typical. Although Typical proved a valuable measure for representing aspects of phenomena that correspond to spread in data, the prevalent place in the report sheets also caused it to dominate the students' choice of measures. The revised design principle instead placed an emphasis on different measures that should be contrasted with each other in order to find a measure suitable for the phenomenon under investigation.

The qualitative analysis of important steps in the learning processes of Kaan & Nesrin and Jana & Mara revealed that the added emphasis on contrasting measures seems to have had the desired effect. Both pairs develop a rich conceptual network encompassing several different situative measures and their represented aspects of the phenomenon. The students no longer focused entirely on Typical, but instead made conscious decisions about the measure to be employed in creating report sheets. Still, Typical served as an important measure for development, as Jana and Mara broadened the contextual neighborhood of their measure Typical.

Most parts of the students' rich reflective activities are seen when contrasting different measures. The influence of the context and the importance of the mathematizing activity of structuring for reflection observed in Cycle III also can be seen again, as both pairs engage in structuring and representing while reflecting. This shows the reflection-initiating function of the Arctic sea ice context. Thus, DP/Context-Revised also shows the desired effects in the two cases of Kaan & Nesrin and Jana & Mara. After the revision of the two design principles, the learning processes seem to have gained in richness of reflective activities and concepts.

The richness of mathematizing activity, however, seems to have lessened. With few exceptions, the students did not engage in the mathematizing activity of formalizing, leaving their measures underdeveloped in terms of formal characteristics and corresponding features of the data. This was true in the analyzed sequences as well as throughout the whole design experiments.

One explanation for this lack of formalizing could be the lacking implementation of the design principle DP/Formalizations. In Cycle III, the students were able to compare different formalizations of the same situative measure of Typical. Variations on Typical included formalizations through a single number, a small interval, and a large interval. This prompted students like Maria and Natalie to identify fixed formal characteristics such as a procedure of calculation.

Although the filled-in report sheets of Cycle IV show a greater variety of measures, they do not show any such variation of formalizations. Every measure employed on the filled-in report sheets of Cycle IV is unique, but because each measure only appeared on a single filled-in report sheet, there was no variety of different formalizations of the same measure to be compared. This seems to have resulted in the students not feeling the need to identify formal characteristics by, for example, specifying the method of calculation of their measures. Also, the prompts given by the design experiment leader had changed, focusing more on justifying the choice of measure (i.e. on DP/Measures-Revised) and less on the exact values and the procedures of calculation for the measures employed (i.e. on DP/Formalizations).

Overall, the changes in design seem to have supported the intended goal of initiating more reflective activities. The students' mathematizing activities, however, do seem to have suffered under these changes. The students still developed situative measures with a pronounced use that is sensitive to different phenomena. Their mathematizing conceptual network also shows a differentiated view of aspects of phenomena. Yet, they remain unfinished in their formal characteristics and corresponding features of the data. A further revision of the design would need to find an approach integrating the design of both design experiment cycles.

8.3 Identification of situative and general reflective concepts

A major goal of the PhD-thesis is to provide a contribution towards closing the specification and realization gaps of the reflective side of mathematical literacy (see Chapters 2 and 4). This can be done by identifying general reflective concepts as reflective goals (regarding the specification gap) and by identifying situative reflective concepts as potential precursors to the development of these general reflective concepts (regarding the realization gap). Although only four case studies were investigated, the analysis already identified a rich repertoire of situative reflective concepts drawn on by the students and possible general reflective concepts that could follow. Table 8.3 gives an overview of the reflective concepts identified in Cycle IV and Cycle III.

Tab. 8.3: Situative and general reflective concepts identified in the study; students' initials are added to their situative concepts

Situative reflective concept	General reflective concept	Description
Correctness (KN)	**Consistency**	Different measures that address the same aspect produce consistent results.
Meaning (KN); Exactness (MN);	**Specificity**	Measures do not address phenomena holistically, but partition phenomena into discrete aspects that then are modelled by measures
Approximate value (QR)	**Summary**	Measures act as summaries of complex data, intentionally sacrificing information to gain efficiency
Understandability (KN, MN)	**Intersubjectivity**	Measures create shared understandings of phenomena by providing objective, shared means of description
What you should say (KN); Importance (JM); Normal winter jacket (MN)	**Contextual relevance**	Although measures can be applied to any phenomenon, their usefulness depends on the phenomenon under investigation

Tab. 8.3: Situative and general reflective concepts identified in the study; students' initials are added to their situative concepts (continued)

Situative reflective concept	General reflective concept	Description
What it looks like (KN); What you can notice (JM); Correct but weird (JM); What it tells you (MN); Rough estimate (MN)	**Perspectivity**	Measures highlight some aspects of phenomena while discarding all others.
What is in your head (KN)	**Formatting power**	Choosing a measure is not a neutral description, but in itself is already an act that can influence the interpretation of a phenomenon; this can be utilized to enforce one's interests
What you can notice (JM); What one can over-look (MN)	**Insight**	Measures can provide information about phenomena that was not available before

Some of these general reflective concepts have already been specified in Chapter 2. The reflective concept of *contextual relevance* is needed for advocating or rejecting experts' advice, and thus is an important concept for the ability to communicate with experts (Fischer, 2001). The reflective concepts of *perspectivity* and *formatting power* can enable individuals to identify the reality-determining power of mathematical modelling (Skovsmose, 2012). These reflective concepts were identified beforehand (see Chapter 2) and used as sensitizing concepts in the analysis (see Chapter 5). However, it was not clear from the beginning in which way seventh graders would be able to articulate ideas corresponding to these concepts. These individual articulations are now captured by the situative reflective concepts.

Other reflective concepts could only be identified through the empirical analyses, but still can be related to general reflective concepts from literature a-posteriori. The reflective concepts of *consistency* and *specificity* relate to the fact that mathematical concepts are situated in consistent conceptual networks, and that mathematical concepts have specific, well-defined areas of applicability. Thus, they concern meta-mathematical knowledge about the way mathematical concepts are used within mathematics (e.g. Lakatos, 1976). The reflective concepts of *summary* and *insight*, on the other hand, refer to meta-knowledge about the practice of mathematics and statistics. Statistical measures summarize data, which enables one to see patterns at the cost of acknowledging variability

(Kröpfl et al., 2000). This can result in new insights into phenomena that could not have been observable otherwise, which is one of the central aims of statistics (Wild & Pfannkuch, 1999). Finally, the reflective concept of *intersubjectivity* refers to a central function of mathematics for society: mathematical concepts provide the objective means of communication which enables materialized and non-personal communication about phenomena (Fischer, 1988).

Thus, a first tentative categorization of reflective concepts can be proposed here: the general reflective concepts specified here belong to the categories of *modelling* (perspectivity, formatting power), *meta-mathematics* (consistency, specificity), *meta-practice* (summary, insights), or *communication* (intersubjectivity, contextual relevance).

Building on this first specification, some observations can be made based on this list of situative and general reflective concepts.

8.3.1 The ambiguity of situative reflective concepts

The situative reflective concepts-in-action were reconstructed as close to students' use of language as possible. This has revealed a variety of situative concepts with very similar articulations: from superficial observation, it is not easy to find differences between *what it is, what it looks like, what you can overlook, what you can notice,* and *what it tells you.* Only the in-depth analysis of the students' reasoning on a micro level has revealed the differences between these articulations, as they can provide precursors to different general reflective concepts. Different students use very similar articulations to refer to different concepts: Kaan and Nesrin use *what it looks like* to refer to perspectivity, whereas Maria and Natalie use *what one can overlook* to refer to *insight.* Jana and Mara again use the single articulation *what you can notice* to refer either to either *perspectivity* or *insight.* Those differences, however, are fuzzy and connections between situative and general reflective concepts remain ambiguous.

This fuzziness of students' situative reflective concepts, however, does not constitute a bad sign for concept development. Regarding mathematizing concepts, it is regarded as common knowledge that when learning, students' initial situative concepts are ambiguous, fuzzy, contextually bound and informally articulated. Yet it also is generally accepted that students can develop general mathematizing concepts from this ambiguous plurality of situative mathematizing concepts. The results presented here show the same ambiguous nature to be true for students' situative reflective concepts, suggesting that with careful guidance, students can indeed develop general reflective concepts.

8.3.2 Three reflective core concepts

Although there are many individual articulations of situative reflective concepts, three general reflective concepts stand out that were addressed by three out of four pairs: *specificity, perspectivity*, and *contextual relevance*. As argued earlier, the existence proof of these three general reflective concepts already is a result worth noting. The existence proof is now extended to general reflective concepts which could be counted as common to the four case studies.

One general reflective concept that could not be identified in the investigated teaching-learning arrangement is the concept of manipulation. This concept is commonly discussed in specifying statistical literacy and refers to intentionally falsifying or omitting statistical information in order to create false views on phenomena (see Chapter 2). Instead, the students drew on the far more powerful reflective concept of perspectivity: unfitting perspectives on phenomena can not only be created by falsified information, but also by any correct use of statistics. Correctness is a meta-mathematical reflective concept - there is no single 'correct' use of statistics creating 'correct' insights on phenomena, but every use of statistics creates a specific perspective on phenomena.

Yet, the general reflective concept of perspectivity so far remains an under-developed objective in the mathematics education discourse. This research suggests that students are reliably able to draw on situative precursors to this general reflective concept. Since the three concepts could be reconstructed so reliably, they are potential candidates to function as *reflective core concepts* for the design of teaching-learning arrangements. These core concepts could provide possible starting points for reflective concept development and a foundation for developing more complex reflective concepts. Further research needs to evaluate how to explicitly incorporate these reflective core concepts into the design of teaching-learning arrangements, and how reflective concept development can be supported.

8.3.3 The possibility and need for development

Besides the reflective core concepts, many other general reflective concepts could be identified during the learning processes. One of these is the concept of *formatting power*, with its situative articulation of *what one reads first*. This is surprising, as it could be argued that the formatting power is such a complex concept that it could never belong to a learning content in school – although Skovsmose (1994) sketches some possible project work for students to address the formatting power of mathematics, empirical accounts of students reflecting on it are still missing. The empirical results presented here, however, show that students can draw on precursors to such a complex concept in their learning processes, although certainly in a situative way.

This situative appearance of the concept, however, also illustrates the need for further concept development. A single appearance of the situative concept of *what is in your head* does not make a fully developed concept of *formatting power*. The reflective goals of mathematical literacy cannot be reached by simply eliciting some situative reflective concepts when convenient, without following through on serious concept development. The task of reflective concept development still needs to be carried out.

This is also illustrated by the case of Kaan and Nesrin. Whereas Nesrin is able to further develop their situative measures by engaging in reflective activities, Kaan seemed to struggle with this task. He was mostly silent in crucial moments of the design experiment, leaving Nesrin the role to reflect on contextual relevance, specificity, and perspectivity. It seems that Kaan did not have access to situative reflective concepts that would have enabled him to do likewise. This illustrates another main point of this thesis: reflection is no natural or intuitive activity to be taken for granted, but rather needs concept development of its own. With the design principle DP/Measures-Refined implemented by means of the filled-in report sheets, the teaching-learning arrangements tried to provide opportunities initiating the development of reflective concepts. With some students, this seemed to work; but others such as Kaan seem to have been left behind. Reasons for this might be that it was never fully explicated what would count as a valid comparison of measures, which left Kaan with little incentive to engage in reflective activity. Other students had little problem with drawing on their situative reflective concepts. Further research will be necessary to find ways to engage all learners in this kind of reflection.

8.3.4 Connections between reflective concepts and activities

The situative reflective concepts reconstructed during the students' learning processes are closely connected to their reflective activities. Table 8.4 shows how the situative concepts were distributed across the different reflective activities.

Tab. 8.4: Situative reflective concepts by student pair and corresponding reflective activity

Student pair	Identifying patterns of thought	Explicating aims and purposes	Denominating risks and limits
M&N	Exactness, What it tells you	Understandability, What one can overlook	What it tells you, Winter Jacket
Q&R	Approximate value		
K&N	Correctness, Meaning	What you should say, What it looks like	Understandability, What is in your head
J&M	Correct but weird	Importance	What you can notice

Some first patterns regarding the connection between reflective concepts and reflective activities can be identified based on Table 8.4. There seems to be a difference between reflective concepts of identifying patterns of thought on the one hand, and reflective concepts of explicating aims and purposes and denominating risks and limits on the other. The former reflective concepts seem to mostly describe 'mathematical' properties: measures being exact, approximate, or correct. The latter seem to mostly refer to communication (understandability, what you should say, what it tells you) and visibility (what one can overlook, what it looks like, what you can notice).

The emerging pattern of different types of reflective concepts also holds when comparing the corresponding general concepts (Tab. 8.5). Some situative 'mathematical' concepts (exactness, correctness) are now substituted by general reflective concepts referring to the category of meta-mathematics (specificity, consistency). These meta-mathematical concepts are mostly used when identifying patterns of thought, whereas the concepts of the other two reflective activities broadly refer to communication (intersubjectivity), meta-practice (insight), or modelling (perspectivity) – although concepts such as perspectivity seem to support all three reflective activities.

Tab. 8.5: General reflective concepts by student pair and corresponding reflective activity

Student pair	Identifying patterns of thought	Explicating aims and purposes	Denominating risks and limits
M&N	Specificity, Perspectivity	Intersubjectivity, Insight	Perspectivity, Contextual Relevance
Q&R	Summary		
K&N	Consistency, Specificity	Contextual Relevance, Perspectivity	Intersubjectivity, Formatting power
J&M	Perspectivity Contextual Relevance	Contextual Relevance	Insight

These observed patterns should only be treated as preliminary, as the number of investigated learning processes remains too small to justify generalizable connections between concepts and activities. However, the preliminary findings do indicate a tendency for the reflective activity of identifying patterns of thought to require reflective concepts different than the other reflective activities.

8.3.5 An existence proof of reflection in learning processes

This list of situative and general reflective concepts is neither intended to be complete nor sufficient for a list of possible concepts as it was induced from only four limited case studies, and bound to the specific teaching learning arrangement. The concepts identified are based on a small number of design experiments and participants, and their articulations are heavily influenced by the concrete learning processes observed. Nevertheless, this list provides an important contribution to the systematic research of students' reflections by providing an *existence proof* of reflection in learning processes. This PhD-thesis does not make the claim that the identified concepts are to be the definitive, consolidated list of situative and general reflective concepts; it does, however, make the claim that students' reflections can and should be the object of systematic research. Additionally, not only does the thesis prove that reflection can take place during learning processes, but also that reflection can be so rich as to address a whole variety of general reflective concepts. Students can and do flexibly draw on a variety of reflective concepts. However, the students observed in this thesis each focused on only few specific reflective concepts. Again, systematic research is needed to map out the possible different situative and reflective concepts as well as conditions and obstacles for students to draw on particular reflective concepts, so that the richness of different reflective concepts can be made available to every student.

Thus, the list of situative reflective concepts only sketches a beginning of research on the development of general reflective concepts. Some possible ways of development have also already been identified. Cycle III showed how some situative reflective concepts remain stable across design experiments; the reflective core concepts could provide a starting point for concept development. However, the nature of this thesis as existence proof of reflection in learning processes allows only little insights into how situative reflective concepts could be developed further. As such, this thesis has only sketched the road ahead for systematic research into students' reflection processes. This concludes the empirical part of this PhD-thesis. The following chapter provides a final summary of the thesis and revisits the central theoretical results as well as the central empirical results.

9 Conclusions

Starting point for this thesis was the need for a compact teaching-learning arrangement that enables students to develop central statistical concepts in a limited number of lessons (Chapter 1). This lead to an investigation into the normative aims of mathematics and statistics education in order to identify the concepts learners need to develop (Chapter 2). Drawing on the framework of mathematical literacy, two sides of mathematical literacy were identified (Section 2.1). The mathematizing side of mathematical literacy concerns the use of mathematics for structuring phenomena of the real world. The reflective side of mathematical literacy concerns the evaluation of the role mathematics has in society. Although grounded in mathematics education research, the activity of mathematizing and the construct of mathematical literacy could also be gainfully employed for statistics education research (Section 2.2). Each of the sides of mathematical literacy was characterized through different general learning aims and specific learning goals. The aims of the mathematizing side were identified as investigating hypothetical situations and describing problems and communicating understandings (Section 2.3.1). The goals of the reflective side were identified as identifying the formatting power of mathematics and evaluating expert judgement (Section 2.4.1). For both sides of mathematical literacy, the goals comprise mastering activities (Sections 2.3.2 and 2.4.2) and developing concepts (Sections 2.3.3 and 2.4.3). But whereas a variety of general mathematizing concepts to be developed could be identified in literature (Section 2.2.3), a specification gap for general reflective concepts was identified (Section 2.3.3). For both sides of mathematical literacy, however, measures were identified as a possible central learning content (Section 2.5).

Chapter 3 specified the learning content of statistical measures. A review of literature revealed some starting points but showed that literature was mostly focused on the mathematizing concept of average and single isolated measures (Section 3.1.1). A conceptualization of measures was introduced through the constructs of formal characteristics, features of data, and aspects of phenomena (Section 3.1.2). A review of empirical studies revealed some insights into learners' possible intuitive measures, yet also the need for further empirical research (Section 3.2). Developing an understanding of measures was identified as a necessary precursor to and a starting point for Informal Statistical Inference (Section 3.3).

Chapter 4 provided the learning-theoretical foundation that emphasized the situativity of knowledge (Chapter 4.1). The Theory of Conceptual Fields was utilized to illustrate the deep connection between learners' concepts and their activities, linking the two goals of mastering activities and developing concepts (Section 4.1.1). The theory of situated abstractions provided a language for the importance of context and the mechanisms of concept development through

© Springer Fachmedien Wiesbaden GmbH, part of Springer Nature 2018
C. Büscher, *Mathematical Literacy on Statistical Measures*,
Dortmunder Beiträge zur Entwicklung und Erforschung des
Mathematikunterrichts 37, https://doi.org/10.1007/978-3-658-23069-2_9

broadening the contextual neighborhood (Section 4.1.2). The two learning-theoretical frameworks provided a language for a hypothetical learning trajectory for students' development of measures for the mathematizing side of mathematical literacy (Section 4.2.1). For the reflective side of mathematical literacy however, a realization gap was identified: missing conceptualizations of learners' situative reflections hinder the formulation of a hypothetical learning trajectory concerning reflective concepts (Section 4.2.2).

This theoretical work led to research questions requiring empirical research (Chapter 5). The underlying empirical study utilized the framework of Topic-specific Didactical Design Research and was illustrated in Chapter 6. Chapters 7 and 8 then presented the design of a compact teaching-learning arrangement as well as empirical analyses.

9.1 Central theoretical results of the thesis

The point of departure for this thesis was the need for a compact teaching-learning arrangement for statistics (i.e. one that achieves the most important goals in a limited number of lessons, see Chapter 1). This was motivated by the fact that for most countries, only very little time is allocated to statistics education within mathematics classrooms (Zieffler et al., 2018). Pursuing the task of designing such a teaching-learning arrangement resulted in central theoretical results. Firstly, a normative framework had to be identified that allowed to specify central learning contents (Section 9.1.1). Secondly, an analytic framework had to identified that allowed to describe learners' processes of concept development at a micro-level (Section 9.1.2). Finally, a descriptive framework hat to be constructed for describing learners' reflections in learning processes (Section 9.1.3).

9.1.1 A framework for specifying the learning content of statistical measures

Chapter 2 drew on the framework of mathematical literacy to specify central learning contents of statistics. Although no clear consensus exists regarding a definition of mathematical literacy (Niss & Jablonka, 2014), two sides of mathematical literacy can generally be distinguished: the *mathematizing side* of mathematical literacy dealing with the ability to use mathematics to describe phenomena (e.g. Freudenthal, 1991). And the *reflective side* of mathematical literacy concerning the ability to evaluate the role mathematics has in the world (Skovsmose, 1994). For each side of mathematical literacy, this thesis identified general learning aims as well as specific learning goals. For the mathematizing side regarding statistics, this resulted in specifying the mathematizing goals as mastering the *mathematizing activities* of structuring, representing, and formalizing, as well as developing the *mathematizing concepts* of distribution, variability, and measure.

Whereas there exist elaborations for the mathematizing concepts of distribution (Wild, 2006) and variability (Gould, 2011), the concept of measure has only been treated indirectly by focusing on single measures of center or spread (e.g. Konold & Pollatsek, 2002; Makar, 2014). Chapter 3 thus provided a general conceptualization of statistical measures through three constructs (Fig. 3.1): *formal characteristics* are mathematical properties like form and procedure of calculation; *features of data* are any identifiable distinct parts of given (such as modal clumps, Konold et al., 2002); *aspects of phenomena* are any identifiably parts of larger phenomena. Measures *represent* aspects of phenomena, which *correspond* to features of data. These then *determine* the concrete value of a measure given some data. Measures were defined as *data-based formalized representations of aspects of phenomena.*

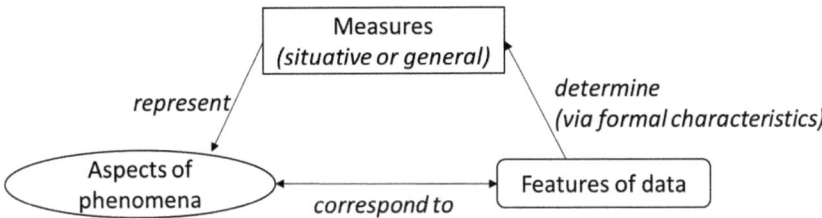

Fig. 3.1: Measures represent aspects of phenomena and are determined by features of data through their formal characteristics

Another distinction was drawn between different types of measures. *General* measures are those measures belonging to the formal body of statistical knowledge. They can be applied to any phenomena, such as mean, median, or standard deviation. In contrast, *situative* measures are bound to specific phenomena, such as the daily ice growth (Fig. 3.2): for the phenomenon of Arctic sea ice, researchers use the daily ice growth to represent the aspect of the melting process of the ice. This aspect corresponds to the feature of the variation observed in the data, which determines the value of the daily ice growth.

Situative measures also comprise any measures used by learners expressing their individual and intuitive knowledge about data, such as their use of modal clumps to represent the typical amount of roadkill for a single day (see Konold et al., 2002). The distinction between general and situative measures could then be used in specifying the prescriptive and analytic framework of concept development.

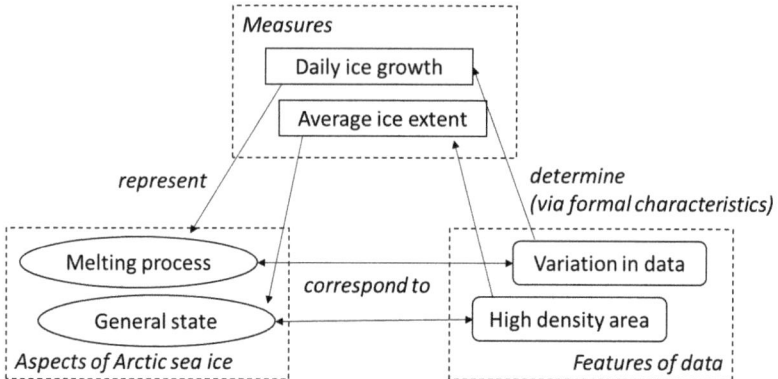

Fig. 3.2: Situative measures of the phenomenon of Arctic sea ice and their represented aspects of the phenomenon and corresponding features of data

9.1.2 A framework for prescribing and analyzing students' processes of developing measures

Statistics education research is in need of research into learning trajectories (Arnold et al., 2018). Although research has identified students' intuitive ways of reasoning (e.g. Konold et al., 2002; Makar & Confrey, 2003; Büscher & Schnell, 2015), more insights are needed into how to foster this informal knowledge into the development of the mathematizing concepts of the mathematizing goals of mathematical literacy such as distribution, variability, and measure. This requires a framework that allows to prescribe students' trajectories of concept development as well as to analyze students' actual learning pathways.

Drawing on the conceptualization of measure, Chapter 4 proposed a hypothetical learning trajectory towards the mathematizing goals of mathematical literacy in statistics (Fig. 4.1). This trajectory is structured among four levels from concrete situations to abstract knowledge (cf. Gravemeijer, 1999). Learners begin by acting in *mathematizing* situations, from which they develop specific *measures*. These measures then become grouped into *classes of measures*, the totality of classes then composing the *mathematizing goal* of the concept of measure.

According to this hypothetical learning trajectory, learners start out by identifying situation-specific aspects of phenomena and features of data. By engaging in the mathematizing activity of *structuring*, they identify new aspects of phenomena, new features of data, or find new connections between aspects of phenomena and their corresponding features of data. They then engage in the mathematizing activity of *representing* to link these aspects of phenomena and features of data to situative or general measures. Measures then get developed further through the mathematizing activity of *formalizing*, which consists in identifying their formal characteristics.

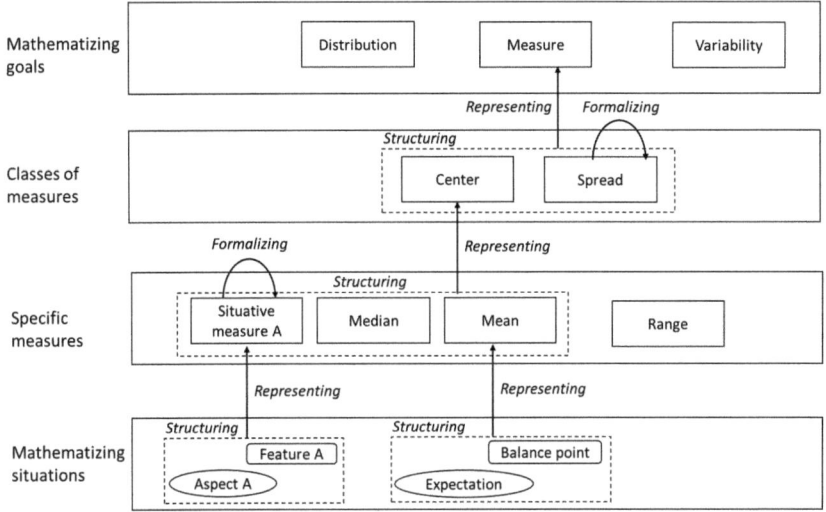

Fig. 4.1: Sketch of a general learning trajectory towards the mathematizing goals

When these measures thus have been specified by learners, they again can be objects of structuring and representing. Similar measures become grouped through the activity of structuring and represented through the higher-order mathematizing concepts of classes of measures (e.g. mean and median both being represented by the class of measures of center). Formalizing these classes of measures can lead to the identification of formal characteristics common to all measures represented by a class. Finally, the concept of measure is developed by representing all classes of measures through a single concept. In this way, a coherent hypothetical learning trajectory leading from students' situative infor-mal thinking up to the highest mathematizing goals through three mathematiz-ing activities is constructed.

The teaching-learning arrangement designed by this thesis mostly focuses on the lower levels of this hypothetical learning trajectory, i.e. the development of specific measures instead of classes of measures. The theoretical constructs of the mathematizing activities and the conceptualization of measure could then be used as an analytical framework for describing students' learning processes. Figure 9.1 presents an excerpt of the learning process one pair of students dur-ing one design experiment as an example (see Chapter 7 for the full analysis). Over the three episodes of the design experiment presented here, the students engage various mathematizing activities. It the beginning, they mostly engage in the mathematizing activity of structuring ('MA/s') by identifying new aspects of phenomena and features of data. During the later parts, they identify several measures (average and Typical) and engage in representing ('MA/r') and formalizing ('MA/f'). In this way, their network of concepts gets increasingly complex while they develop their measures.

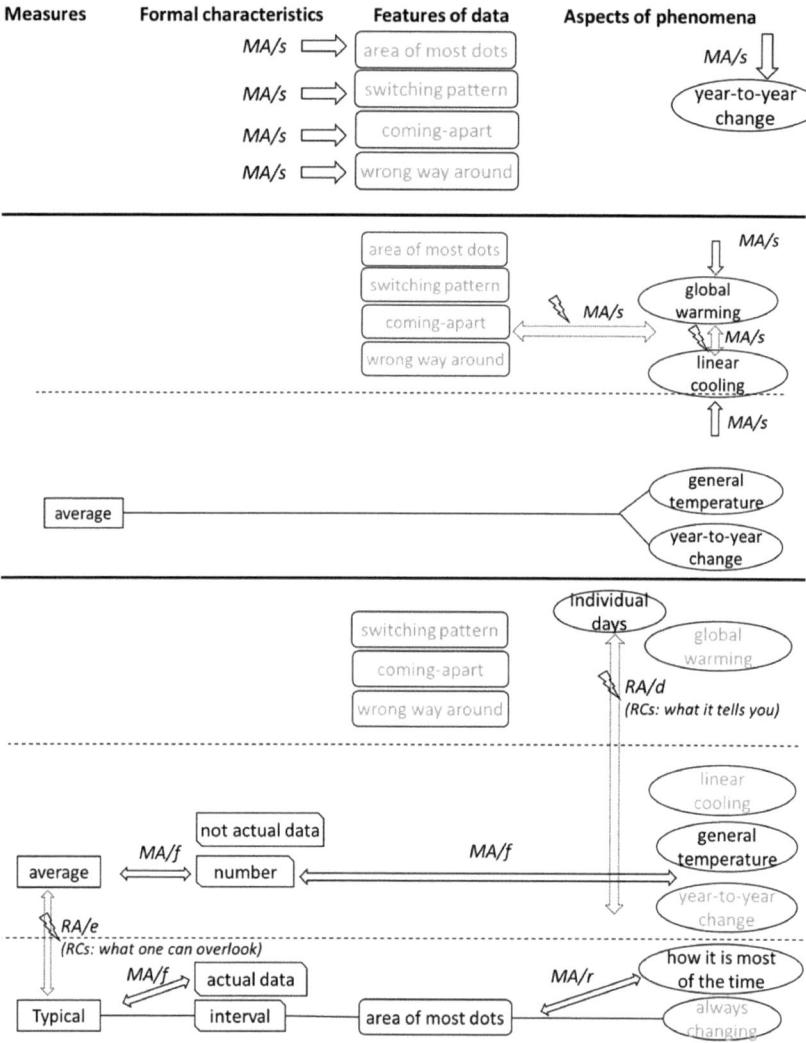

Fig. 9.1: Excerpt of Maria and Natalie's learning process during the Typical Antarctic Temperatures Problem

This investigation of learning processes allowed this thesis to find differences and similarities in the students' learning processes in order to identify typical pathways and obstacles. This theoretical result of a framework for describing learning processes thus laid the groundwork for the empirical results of this thesis (Section 9.2).

9.1.3 A descriptive framework of reflection as a conceptual activity

The general need for students to be able to reflect on mathematics has a long-standing tradition in mathematics education research (e.g. Skovsmose, 1994; Wille, 1995). Although some proposals for bringing reflection into the class-room do exist (e.g. Skovsmose, 1998; Lengnink, 2010), detailed empirical accounts of students' reflections were found to be in limited supply.

This thesis introduced the constructs of reflective activities and reflective concepts. Drawing on ideas proposed by Wille (1995), this thesis conceptualized reflection as consisting of the reflective activities of *identifying patterns of thought, explicating aims and purposes,* and *denominating risks and limits* (Chapter 2). Utilizing the learning-theoretical Theory of Conceptual Fields (Vergnaud, 1996), this thesis treated these reflective activities not like impalpable higher-order thinking, but like any other activity: as organizing behavior shaped by concepts (Chapter 4). Just as mathematizing activities draw on mathematizing concepts, so reflective activities draw on reflective concepts. And just as mathematizing concepts can take the form on individual and situative concepts, so can reflective concepts. This opens up the possibility of developing students' reflective concepts via teaching-learning arrangements focused on initiating reflective activities – and also enables the identification of the specification and realization gaps for reflective concepts. However, due to the identified realization gap of reflective concepts, the actual identification of situative reflective concepts remained as an empirical task. These theoretical considerations could only later be grounded empirically (Section 9.2.4). Thus, the construct-pair of reflective activities and reflective concepts constitutes another major theoretical result of this thesis, as the descriptive framework later enabled the identification of students' reflections during their learning processes.

9.2 Central empirical results of this thesis

After the theoretical work, Chapter 5 introduced the three empirical research questions. The first research question concerned the mathematizing side of mathematical literacy, the second the reflective side of mathematical literacy, and the third the interplay between mathematizing and reflective activities. One major result of Chapters 7 and 8, however, is how both sides of mathematical literacy are intertwined in the development of measures. Therefore, this thesis does not hold up the analytic distinction between both sides of mathematical literacy, but rather answers its empirical research questions in an integrated fashion. The first result is given through the design principle for a compact teaching-learning arrangement for both sides of mathematical literacy (Section 9.2.1). Special emphasis is given to the context used in a teaching-learning arrangement (Section 9.2.2). The third result concerns students' development of measures through the interplay of mathematizing and reflective activities (Section 9.2.3). The fourth and final result again focuses on the reflective side of mathematical literacy by providing a contribution to closing the gaps of specification and realization for the reflective side of mathematical literacy (Section 9.2.4). This section concludes with a final summary about all central theoretical as well as empirical results (9.2.5).

9.2.1 Design of a compact teaching-learning arrangement and design principles for the development of measures

The motivating purpose of this thesis was the need for a compact teaching-learning arrangement for statistics applicable in Grade 7 German classrooms (Chapter 1). This prompted an investigation into the aims and goals of mathematics education, leading to the learning content of statistical measures (Chapter 2). Statistical measures in general, however, seem to have rarely been the focus of research in statistics education (Chapter 3). Thus, in order to enact the hypothetical learning trajectory (Chapter 4), this thesis identified first design principles for the development of measures (Section 7.1.1). These design principles were then implemented in a teaching-learning arrangement and refined based on the empirical analysis (Section 8.1.1). This resulted in three theoretically and empirically grounded design principles:

- *To enable concept-development based on learners' situative mathematizing and reflective concepts, choose a phenomenon familiar to the students for the context of the problem, because this can enable them to engage in the mathematizing activity of structuring as well as in various reflective activities ('DP/Context-Refined').*
- *To increase the conceptual richness of learners' mathematizing and reflective activities, put them into a position that requires them to contrast different measures, because contrasting measures can prompt further mathematizing activities of structuring and representing as well as reflective activities of explicating aims and purposes and denominating risks and limits ('DP/Measures-Refined').*
- *To support the identification of formal characteristics, explicitly call for and challenge students formalizations, because this can prompt further mathematizing activity of formalizing ('DP/Formalizations').*

These design principles were implemented in the Antarctic Temperatures Problem and Arctic Sea Ice Problem (Sections 8.1.1 and 8.1.2). These problems asked students to predict the behavior of the phenomena of Antarctic weather and Artic sea ice, and to give short summarizing reports. Since problem design consisted of a complex interplay of data, didactic material, and prompts by the design experiment leader, only the central design element of the problems is illustrated here (for full descriptions see Chapters 7 and 8). Both problems made central use of 'report sheets' (Fig. 8.1). These report sheets confronted the students with different interpretations and formalizations of situative measures (e.g. 'Typical' and 'Most Important Value'). The students were asked to create their own report sheets, using a situative measure of their choice. For this, they created their own measures and had to justify their choice. Both problems proved fruitful, as the students engaged in various mathematizing activities supporting the development of their situative measures.

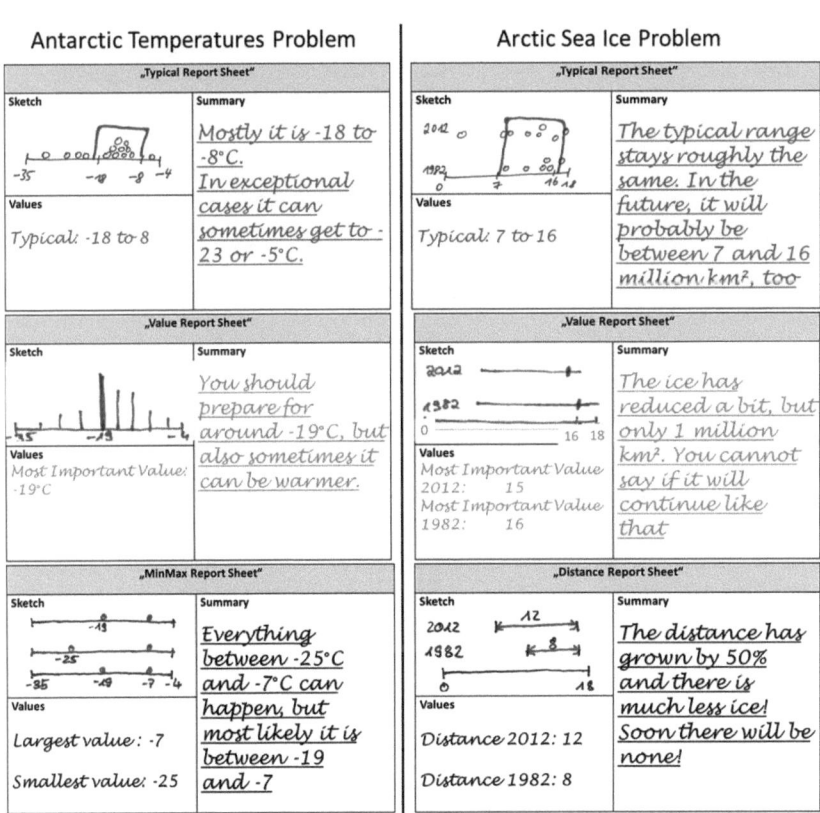

Fig. 8.1: The filled-in report sheets for Cycle IV

Two cycles of design research were conducted with similar teaching-learning arrangements. However, due to changes to the report sheets, the implementations of the design principles differed slightly. One variant of the teaching-learning arrangement focused on design principle DP/Formalizations (Chapter 7), whereas the other placed less emphasis on this design principle (Chapter 8). This resulted in more pronounced mathematizing activities of formalizing for the former variant, whereas the latter variant focused on reflective activities.

Thus, the second variant of the teaching-learning arrangement (Chapter 8) cannot be counted as a direct improvement over the first variant (Chapter 7). The variants differ in the type of activity in focus. Therefore, this thesis has not fully achieved its goal of developing a single compact teaching-learning arrangement, but rather has developed two variants of a teaching-learning arrangement. Further design research needs to consolidate the two variants.

9.2.2 The epistemic and reflection-initiating functions of the Antarctic weather and Arctic sea ice contexts

The first research question asked how a teaching-learning arrangement can support the mathematizing goals of mathematical literacy in statistics, and specifically what a suitable context for a teaching-learning arrangement is. Such a context allows learners to develop situative measures through mathematizing activities (Chapter 5). The design principle DP/Context demanded to choose a context that fulfills not only a motivating or empowering function, but also an epistemic function (Chapter 7): by enabling students to draw on their intuitive situative mathematizing concepts, such a context can directly support students' concept development. This epistemic function of context was already thoroughly utilized by RME Theory for learning contents such as numbers, fractions, and percentages (e.g. van den Heuvel-Panhuizen, 2003; Gravemeijer, 1999). Such work, however, was still largely underdeveloped for the learning content of statistical measures (an exception being the work of Bakker, 2004.

The contexts of Antarctic weather and Arctic sea ice utilized in the design of the teaching-learning arrangement proved to fulfill such an epistemic function in the empirical analysis (Chapters 7 and 8). The aspects of phenomena identified by the students provided the basis for their development of measures. By drawing on their pre-understandings of weather and climate, the students used their situative mathematizing concepts to compare and contrast different measures. The students actively expanded the contextual neighborhood of their measures. In some cases, they drew on their knowledge of Antarctic weather in order to articulate their views on Arctic sea ice.

The empirical analysis also showed that, although in rare cases students focused entirely on features of data, identifying aspects of phenomena took precedence to identifying features of data. For data on Antarctic temperatures, the students focused on features of data like the area of most dots or modal clumps (similar to Konold & Pollatsek, 2002). For data on Arctic sea ice, however, they recognized the modal clumps, but instead chose to focus the minimum or the total spread of data (see Chapter 7). Thus, students do not seem to simply intuitively focus on specific features of data like modal clumps (as proposed by Konold & Pollatsek, 2002), but rather do so according to other criteria.

This behavior can be explained by acknowledging the aspects of phenomena identified by the students as well as the measures they prominently used. For Antarctic temperatures, the modal clumps of the data corresponded to the aspect of Antarctic temperatures of 'likely' or 'main' temperatures. Regarding measures, the students focused on the situative measure 'Typical', which they used to represent these aspects of the phenomenon. For Arctic sea ice, the overall spread of the data corresponded to the aspect of the melting Arctic ice, represented by the situative measure 'Distance'. Thus, in both cases, the choice of measure to be used influenced the students' focus on features of data.

Finally, this can be explained by taking students' reflective activities into account. The students justified their choice of measures by drawing on the reflective concept of contextual relevance. For Arctic sea ice, they did not focus on the feature of modal clumps, because the corresponding aspect of the 'high point' of ice was not held to be relevant in the context by the students. Instead, the feature of overall spread corresponded to the aspect of melting ice, which was held to be much more relevant. Their choice of features of data to be focused on thus was guided by their identified aspects of the phenomenon and perceived contextual relevance.

This shows how students' reasoning cannot adequately be explained by reconstructing which features of data they focus on alone. The analysis of their mathematizing and reflective activities revealed a network of concepts far richer than a simple look at their identified features of data would have revealed. Again, this shows the effects of the epistemic function of the Antarctic weather and Arctic sea ice contexts.

In addition to the epistemic function of context, another function could be identified by this thesis. The students' reflective activities were heavily influenced by their context knowledge, and often coincided with mathematizing activities of structuring. The contexts were central to their reflective activities, and thus were found to be also fulfilling a *reflection-initiating* function. The identification of this function led to the revised design principles DP/Context-Revised (Chapter 8).

9.2.3 Developing measures through mathematizing and reflective activities

Previous research in statistics education has indicated that learners focus on various features of data like modal clumps (Konold et al., 2002), chunks of data (Makar & Confrey, 2003), or case and frequency values (Konold et al., 2015).

All of these studies call for teaching-learning arrangements to build on such learners' reasoning; this thesis provides a contribution to this call. For the students' development of measures, the mathematizing activity of structuring proved to be the central activity, often accompanying other activities. By engaging in various mathematizing activities, the students were able to develop their situative measures of Typical, belonging to sophisticated networks of formal characteristics, features of data, and aspects of the phenomenon. The common language of empirical analysis and specification of learning content revealed the similarities between students' situative measures and general measures (Tab. 7.27). The formal characteristics, corresponding features of data, and represented aspects of phenomena of the situative measure 'Typical' developed by two pairs of students appeared as situative variants of the general measure of interquartile range. This shows how general mathematizing concepts can be developed from the students' situative concepts.

Tab. 7.27: Maria and Natalie's (M&N) and Quanna and Rebecca's (Q&R) respective situative measures 'Typical' show similarities to the general measure 'interquartile range'

Measure	Characteristics	Features	Aspects
Typical (M&N)	Interval, Actual data, Calculation through average	Area of most dots	How it is most of the time, Always varying, Single days
Typical (Q&R)	Interval, Not more than half, Middle of interval	Area of most dots	Likely recurring temperatures, Varying temperatures, Changes in Arctic sea ice
Interquartile range (general concept)	Length of interval, Middle 50% of data, Calculation through median	Dense area in middle of data	Main area of data, Expected deviation from expectation

The process of developing measures proved to be complex. Aside from often starting out with the activity of structuring, no hierarchical model of activities could be reconstructed; instead, analysis illustrated a complex mix of interchanging and interconnected activities.

The complexity was increased further by the fact that mathematizing and reflective activity were intertwined. Although some authors do show how reflection can take place within mathematics classroom practice (e.g. Prediger, 2005b; Kröpfl, Peschek, & Schneider, 2010), reflection is commonly treated as distinct from developing mathematizing concepts. Contrary to that position, the empirical analysis showed that the most crucial parts of mathematizing concept development often saw accompanying reflective activity; and when reflective activities were missing, the mathematizing activities were often found lacking. On the other hand, the aspects of phenomena and formal characteristics resulting from mathematizing activity firstly provided the objects of reflective activity. This shows how mathematizing and reflective activities can reciprocally support one another. Thus, a thorough description of learners' mathematizing concept development will also need to pay attention to their reflective activities and concepts.

9.2.4 Reducing the specification and realization gaps for reflective concepts

During the analysis, several situative reflective concepts shown by the learners have been reconstructed, along with general reflective concepts they could be precursors to (Tab. 8.3).

Tab. 8.3:   Situative and general reflective concepts identified in the study; students' initials are added to their situative concepts

Situative reflective concept	General reflective concept	Description
Correctness (KN)	**Consistency**	Different measures that address the same aspect produce consistent results.
Meaning (KN); Exactness (MN);	**Specificity**	Measures do not address phenomena holistically, but partition phenomena into discrete aspects that then are modelled by measures
Approximate value (QR)	**Summary**	Measures act as summaries of complex data, intentionally sacrificing information to gain efficiency
Understandability (KN, MN)	**Intersubjectivity**	Measures create shared understandings of phenomena by providing objective, shared means of description
What you should say (KN); Importance (JM); Normal winter jacket (MN)	**Contextual relevance**	Although measures can be applied to any phenomenon, their usefulness depends on the phenomenon under investigation
What it looks like (KN); What you can notice (JM); Correct but weird (JM); What it tells you (MN); Rough estimate (MN)	**Perspectivity**	Measures highlight some aspects of phenomena while discarding all others.

Tab. 8.3:   Situative and general reflective concepts identified in the study; students' initials
are added to their situative concepts (continued)

Situative reflective concept	General reflective concept	Description
What is in your head (KN)	**Formatting power**	Choosing a measure is not a neutral description, but in itself is already an act that can influence the interpretation of a phenomenon; this can be utilized to enforce one's interests
What you can notice (JM); What one can overlook (MN)	**Insight**	Measures can provide information about phenomena that was not available before

Just like situative mathematizing concepts, the situative reflective concepts
proved to be fuzzy, unstable, and with unclear boundaries. Yet they played an
important part in supporting the students' reflective activities, and some of the
students' situative reflective concepts even remained stable across design experiments. This revealed how in some cases, students can even address complex
general reflective concepts like the concept of formatting power – in an individual, situative way. It thus seems possible to initiate reflective activities and to
appreciate students' situative reflective concepts as what they are: stepping
stones towards the development of general reflective concepts and the attainment of the reflective goals of mathematical literacy.

Work thus still needs to be done to build on these results in further systematic research. With these and additional connecting points for further research in
mind, this thesis turns towards its limitations and necessary further research.

9.3 Limitations and necessary further research

Denominating risks and limits is not only a worthwhile activity for the development of measures, but also for further development of mathematics education
theory and empirical research. Several possible research studies could be carried
out building on this thesis.

9.3.1 The local nature of design research

The methodological framework of Design Research is optimized for achieving
deep insights into learning processes, aiming for a comprehensive description
and explanation of the processes involved. The resulting theories, however,
inevitably remain local, and careful attention has to be given to the generalizability of results (Prediger et al., 2015).

The local nature of design research can have different articulations. The
Design Research study remains *local in scope*. Only one teaching-learning

arrangement was developed, utilizing two different phenomena as context. This thesis can support no claims on whether another context or other design elements could have had even more beneficial effects. Aside from this, the limited scope also concerns this thesis' generalizability: In total, 34 students in 32 design experiments were investigated. Although the analyzed video material consisted of 855 minutes of video data and 170 pages of transcripts, this is a too small data corpus for statistical generalizability even for the observed students' classes. This thesis, however, does not claim such statistical validity. The sampling employed by the study also does not intend to cover a wide range of participants. Due to the exploratory nature of this thesis, it instead aims at ecological validity by giving a thorough account of the complex learning processes of the students involved. The general theoretical framework provides some grounds for claims of generalizability of results; it is a task of quantitative research to investigate. A possible quantitative follow-up research could take the form of an intervention study to provide evidence for efficacy. Such a study could compare experimental groups of classrooms implementing the design principles with control groups of traditional statistics education. The dependent variable measured with respect to learning gains could be assessed by an adapted test of statistical literacy for German classrooms (e.g. Callingham & Watson, 2017). However, as the developed teaching-learning arrangement still needs refinement before it can be used in classrooms, such quantitative research was not yet feasible at this point.

Another limitation is that the Design Research study remains *local in place*. The design experiments of this project were conducted with students of four different schools in one federal state in Germany. It is possible, even likely, that a different educational or cultural background would have an impact on students' situative measures and their development. For example, it could be possible that students living in Arctic regions could find the Antarctic weather context to be unrealistic. Additional qualitative research could use an ethnomathematical approach (Bishop, 1988) to provide more insights into cultural and educational preconditions of concept development and uncover implicit assumptions held by this thesis.

This thesis also remains *local in content*. The Design Research study investigated the development of measures through a single teaching-learning arrangement. Owing to the situative nature of concepts, the mathematizing concepts as well as reflective concepts observed in the study were heavily influenced by this teaching-learning arrangement. Differences in the design of the teaching-learning arrangement would likely lead to differences in initiated activities and observed concepts. Additionally, the study focused on measures belonging to the classes of center and spread. Other measures might not easily be integrated into the theoretical language surrounding aspects of phenomena. Other mathematizing concepts of mathematizing goals, such as variability and sampling, might require additional terminology. The construction of local theories of learning concepts other than the measures investigated here is a task for additional design research.

Finally, although not specific to design research, this thesis remains *local in theory*. It has drawn on background and foreground theories exhibiting a constructivist approach to learning, focusing on individual activity and framed in a mathematics-didactical normative framework. Even more implicit assumptions are likely to have guided this research, which could only be uncovered by comparison with theories of other background (Prediger, 2015). One possible direction for further research could be the assumptions regarding the form of knowledge. Following Vergnaud (1996), this thesis conceptualizes knowledge as concepts held by the individual and implicitly or explicitly activated in activities. Other background theories locate knowledge at the level of interaction between participants of discourse. The theory of commognition (Sfard, 2008) could provide insights into how knowledge is constructed on this level. Such a study would then use the theory of commognition for analyzing the same data in order to compare and contrast the likely different results produced by different theoretical backgrounds.

9.3.2 The nature of existence proof for situative reflective concepts

The identification of situative and general reflective concepts mostly presents an existence proof of reflection during learning processes (see Section 8.3.5): Owing to the local nature of design research, the list of reflective concepts can neither be counted as complete nor sufficient. It also remains unclear how general reflective concepts can be developed from students' situative reflective concepts. However, as argued earlier, it was first necessary to develop a theory of learners' situative reflective activities before questions concerning the development of reflective concepts can be addressed. As such, this thesis has laid the groundwork for a systematic research program into the development of reflective concepts.

The remaining tasks are numerous and suited to different kinds of research. The development of situative into general reflective concepts remains unclear; a task for further design research. Since design research is local in content, different reflective concepts might become important for different learning contents, which could be uncovered by qualitative research. Quantitative research in the form of field studies could provide insights into reflective concept commonly held by students, adding to the consolidation of a list of situative reflective concepts. Finally, the general reflective concepts identified here resulted from empirical reconstruction, and their links to the reflective goals remain ambiguous. Theoretical or philosophical research is needed to integrate them into a larger normative framework.

9.3.3 Adapting for teaching practice

The design experiments took place in a laboratory setting with pairs of students. As such, work remains concerning the implementation of results into teaching practice. Further development needs to create tasks suitable for whole classroom settings, possibly in need of the identification of additional phenomena apart from temperatures and sea ice. The design principles can provide an orientation for task design and teacher prompts. Some results of this thesis have already been implemented into a text book, adapting the use of report sheets for the classroom (Barzel, Prediger, Hußmann, & Leuders, 2017).

This thesis also shows the importance of appreciating students' situative reflective concepts. Integrating reflection into learning processes, however, calls for deeper changes into teaching practice than can be achieved through development of additional tasks alone. Results need to be adapted for large-scale professional development. Further research is necessary to reveal what teachers and what professional developers need to know (Prediger, Leuders, & Rösken-Winter, 2017). This is a task for design research with a focus on professional development (Prediger, Rösike, & Schnell, 2016).

9.4 Summary and outlook

Statistics education needs to provide contributions to both, the mathematizing and the reflective side of mathematical literacy. This means that students need to develop mathematizing and reflective concepts and to master mathematizing and reflective activities. This is a complex task, that only gets more complex by the limited time allocated to statistics education within mathematics classrooms around the world.

The concept of measure provides a learning content that can contribute to both sides of mathematical literacy. Moreover, it is a learning content that can draw on learners' intuitive reasoning through their situative mathematizing concepts and that can lead to the highest mathematizing goals of mathematical literacy. The empirical results of this thesis already show that through careful design of a teaching-learning arrangement, learners' situative measures can resemble general measures after only two design experiment sessions. Additionally, the students engaged in mathematizing and reflective activities simultaneously. Their reflective concepts, although situative, could be identified as precursors to important general reflective concepts that normally would not be part of 7th Grade mathematics education. Their reflective activities supported their mathematizing activities, and vice versa. Thus, the reflective side of mathematical literacy does not present a supplemental part of becoming mathematically literate. Instead, it can (and should!) be integrated into the development of the mathematizing side of mathematical literacy.

Although statistics education can be a challenging task pursuing many different aims with very limited resources, this thesis hopes to have shown that through careful design of teaching-learning arrangements, this complex task can become a manageable one. Further work should be carried out for enabling students to become mathematically literate on statistical measures.

References

Ainley, J., Pratt, D., & Hansen, A. (2006). Connecting Engagement and Focus in Pedagogic Task Design. *British Educational Research Journal, 32*(1), 23–38. https://doi.org/10.1080/01411920500401971

Alrø, H., & Skovsmose, O. (2003). *Dialogue and Learning in Mathematics Education: Intention, Reflection, Critique.* Dordrecht: Kluwer.

Aridor, K., & Ben-Zvi, D. (2017). The Co-Emergence of Aggregate and Modelling Reasoning. *Statistics Education Research Journal, 16*(2), 38–63.

Arnold, P., Confrey, J., Jones, R. S., Lee, H. S., & Pfannkuch, M. (2018). Statistics Learning Trajectories. In D. Ben-Zvi, K. Makar, & J. Garfield (Eds.), *International Handbook of Research in Statistics Education* (pp. 295–326). Cham: Springer.

ASA, GAISE College Report ASA Revision Committee. (2016). *Guidelines for Assessment and Instruction in Statistics Education College Report 2016.* Retrieved from http://www.amstat.org/education/gaise

Bakker, A. (2004). *Design Research in Statistics Education: On Symbolizing and Computer Tools.* Utrecht: CD-β Press.

Bakker, A., & Derry, J. (2011). Lessons from Inferentialism for Statistics Education. *Mathematical Thinking and Learning, 13*(1-2), 5–26.

Bakker, A., & Gravemeijer, K. P. E. (2004). Learning to Reason About Distribution. In D. Ben-Zvi & J. Garfield (Eds.), *The Challenge of Developing Statistical Literacy, Reasoning and Thinking* (pp. 147–168). Dordrecht: Springer. doi:10.1007/1-4020-2278-6_7

Bakker, A., & Gravemeijer, K. P. E. (2006). An Historical Phenomenology of Mean and Median. *Educational Studies in Mathematics, 62*(2), 149–168. https://doi.org/10.1007/s10649-006-7099-8

Barwell, R. (2013). The mathematical formatting of climate change: Critical mathematics education and post-normal science. *Research in Mathematics Education, 15*(1), 1–16. doi:10.1080/14794802.2012.756633

Barzel, B., Prediger, S., Hußmann, S., & Leuders, T. (Eds.). (2017). *Mathewerkstatt.* Berlin: Cornelsen.

Ben-Zvi, D. (2004). Reasoning about variability in comparing distributions. *Statistics Education Research Journal, 3*(2), 42–63.

Ben-Zvi, D., & Arcavi, A. (2001). Junior high school students' construction of global views of data and data representation. *Educational Studies in Mathematics, 45*(1/3), 35–65. https://doi.org/10.1023/A:1013809201228

Ben-Zvi, D., Bakker, A., & Makar, K. (2015). Learning to reason from samples. *Educational Studies in Mathematics, 88*(3), 291–303. https://doi.org/10.1007/s10649-015-9593-3

Ben-Zvi, D., & Makar, K. (Eds.). (2016). *The Teaching and Learning of Statistics: International Perspectives.* Cham: Springer.

Ben-Zvi, D., Gravemeijer, K., & Ainley, J. (2018). Design of Statistics Learning Environments. In D. Ben-Zvi, K. Makar, & J. Garfield (Eds.), *International Handbook of Research in Statistics Education* (pp. 473–502). Cham: Springer.

© Springer Fachmedien Wiesbaden GmbH, part of Springer Nature 2018
C. Büscher, *Mathematical Literacy on Statistical Measures,*
Dortmunder Beiträge zur Entwicklung und Erforschung des
Mathematikunterrichts 37, https://doi.org/10.1007/978-3-658-23069-2

Biehler, R., Ben-Zvi, D., Bakker, A., & Makar, K. (2013). Technology for Enhancing Statistical Reasoning at the School Level. In M. A. Clements, A. J. Bishop, C. Keitel, J. Kilpatrick, & F. K. Leung (Eds.), *Springer International Handbooks of Education: Vol. 27. Third International Handbook of Mathematics Education* (pp. 643–689). New York, NY: Springer. doi:10.1007/978-1-4614-4684-2_21

Bishop, A. J. (1988). Mathematics education in its cultural context. *Educational Studies in Mathematics, 19*(2), 179–191. https://doi.org/10.1007/BF00751231

Bohnsack, R. (2008). The Interpretation of Pictures and the Documentary Method. *Forum Qualitative Social Sciences, 9*(3). https://doi.org/10.17169/fqs-9.3.1171

Brown, A. L. (1992). Design Experiments: Theoretical and Methodological Challenges in Creating Complex Interventions in Classroom Settings. *Journal of the Learning Sciences, 2*(2), 141–178. https://doi.org/10.1207/s15327809jls0202_2

Brown, J. S., Collins, A., & Duguid, P. (1989). Situated Cognition and the Culture of Learning. *Educational Researcher, 18*(1), 32–42.

Burrill, G., & Biehler, R. (2011). Fundamental Statistical Ideas in the School Curriculum and in Training Teachers. In C. Batanero, G. Burrill, & C. Reading (Eds.), *Teaching Statistics in School Mathematics Challenges for Teaching and Teacher Education. A Joint ICMI/IASE Study: The 18th ICMI Study*. Dordrecht: Springer.

Büscher, C. (2015). Was ist normal? – Individuelle Konzepte von Normalität als Fundament für den Vorstellungsaufbau in der Statistik. In F. Caluori, h. Linneweber-Lammerskitten, & C. Streit (Eds.), *Beiträge zum Mathematikunterricht 2015* (pp. 224–227). Münster: WTM.

Büscher, C. (2017). Common Patterns of Thought and Statistics: Accessing Variability Through the Typical. In T. Dooley & G. Gueudet (Eds.), *Proceedings of the Tenth Congress of the European Society for Research in Mathematics Education* (pp. 716–723). Dublin: DCU Institute of Education and ERME.

Büscher, C. (2018, in press). Students' Development of Measures. In G. Burrill & D. Ben-Zvi (Eds.), *Teaching and Learning Statistics: International Perspectives (Volume II)*. Dordrecht: Springer.

Büscher, C. (submitted). *Clumps or Chunks? - Contextual Relevance of Students' Features of the Data*. Paper submitted for presentation at the 42nd Annual Meeting of the International Group for the Psychology of Mathematics Education, Umea, Sweden.

Büscher, C., & Schnell, S. (2017). Students' Emergent Modelling of Statistical Measures - A Case Study. *Statistics Education Research Journal, 16*(2), 144–162.

Büscher, C., & Prediger, S. (submitted). Students' reflection processes on and about statistical measures: A design research study on potential contributions to mathematical literacy. *Journal für Mathematik-Didaktik*.

Callingham, R., & Watson, J. (2017). The Development of Statistical Literacy at School. *Statistics Education Research Journal, 16*(1), 181–201.

Common Core Standards Initiative (CCSSI). (2018, January). Mathematics standards. Retrieved from http://www.corestandards.com/Math/

Corbin, J. M., & Strauss, A. (1990). Grounded theory research: Procedures, canons, and evaluative criteria. *Qualitative Sociology, 13*(1), 3–21. https://doi.org/10.1007/BF00988593

Chovanetz, C., & Schneider, E. (2008). Einer für alle, alle für einen - Reflektieren über Konzepte und Ideen der Beschreibenden Statistik. *PM - Praxis der Mathematik in der Schule, 50*(20), 12–18.

Cobb, G. W., & Moore, D. S. (1997). Mathematics, Statistics, and Teaching. *The American Mathematical Monthly, 104*(9), 801. https://doi.org/10.2307/2975286

Cobb, P., Confrey, J., diSessa, A., Lehrer, R., & Schauble, L. (2003). Design Experiments in Educational Research. *Educational Researcher, 32*(1), 9–13.

de Lange, J. (1997). Using and Applying Mathematics in Education. In A. J. Bishop, K. Clements, C. Keitel, J. Kilpatrick, & C. Laborde (Eds.), *International Handbook of Mathematics Education* (pp. 49–97). Dordrecht: Springer. https://doi.org/10.1007/978-94-009-1465-0_3

del Mas, R. C. (2004). A Comparison of Mathematical and Statistical Reasoning. In D. Ben-Zvi & J. Garfield (Eds.), *The Challenge of Developing Statistical Literacy, Reasoning and Thinking* (pp. 79–95). Dordrecht: Springer. https://doi.org/10.1007/1-4020-2278-6_4

diSessa, A., & Cobb, P. (2004). Ontological Innovation and the Role of Theory in Design Experiments. *The Journal of the Learning Sciences, 13*(1), 77–103.

Duit, R., Gropengießer, H., Kattmann, U., Komorek, M., & Parchmann, I. (2012). The Model of Educational Reconstruction – a Framework for Improving Teaching and Learning Science1. In D. Jorde & J. Dillon (Eds.), *Science Education Research and Practice in Europe* (pp. 13–37). Rotterdam: Sense.

Engel, J. (2017). Statistical Literacy for Active Citizenship: A Call for Data Science Education. *Statistics Education Research Journal, 16*(2), 44–49.

Fetterer, F., Knowles, K., Meier, W., & Savoie, M. (2002, updated daily). *Sea ice index, version 1: Arctic sea ice extent.* NSIDC: National Snow and Ice Data Center. http://dx.doi.org/10.7265/N5QJ7F7W

Feyerabend, P. (1978). *Science in a free society.* London: NLB.

Fischbein, E. (1999). Intuitions and Schemata in Mathematical Reasoning. *Educational Studies in Mathematics, 38*(1/3), 11–50. https://doi.org/10.1023/A:1003488222875

Fischer, R. (1986). Zum Verhältnis von Mathematik und Kommunikation. *mathematica didactica, 9*(3/4), 119–131.

Fischer, R. (1988). Didactics, Mathematics, and Communication. *For the Learning of Mathematics, 8*(2), 20–30.

Fischer, R. (1993a). Mathematics as a means and as a system. In S. P. Restivo, J. P. van Bendegem, & R. Fischer (Eds.), *SUNY series in science, technology, and society. Math worlds. Philosophical and social studies of mathematics and mathematics education.* Albany: State University of New York Press

Fischer, R. (1993b). Mathematics and Social Change. In S. P. Restivo, J. P. van Bendegem, & R. Fischer (Eds.), *SUNY series in science, technology, and society. Math worlds. Philosophical and social studies of mathematics and mathematics education.* Albany: State University of New York Press.

Fischer, R. (2001). Höhere Allgemeinbildung. In R. Aulke (Ed.), *Franz-Fischer-Jahrbücher: Vol. 6. Situation - Ursprung der Bildung.* Norderstedt: Fischer.

Fischer, R. (2013). Entscheidungs-Bildung und Mathematik. In M. Rathgeb, M. Helmerich, R. Krömer, K. Lengnink, & G. Nickel (Eds.), *Mathematik im Prozess. Philosophische, Historische und Didaktische Perspektiven* (pp. 335–345). Dordrecht: Springer. https://doi.org/10.1007/978-3-658-02274-7_24

Fischer, R., & Malle, G. (2004). *Mensch und Mathematik: Eine Einführung in didaktisches Denken und Handeln.* München: Profil.

Freudenthal, H. (1973). *Mathematics as an educational task.* Dordrecht: Reidel.

Freudenthal, H. (1983a). *Didactical Phenomenology of Mathematical Structures.* Dordrecht: Reidel.

Freudenthal, H. (1983b). Wie entwickelt sich reflexives Denken? *Neue Sammlung, 23*(5), 485–497.

Freudenthal, H. (1991). *Revisiting Mathematics Education: China Lectures.* Dordrecht: Kluwer.

Frischemeier, D. (2017). *Statistisch denken und forschen lernen mit der Software TinkerPlots.* Wiesbaden: Springer Spektrum.

Gal, I. (2002). Adults' Statistical Literacy: Meanings, Components, Responsibilities. *International Statistical Review, 70*(1), 1–25.

Garfield, J., Le, L., Zieffler, A., & Ben-Zvi, D. (2015). Developing students' reasoning about samples and sampling variability as a path to expert statistical thinking. *Educational Studies in Mathematics, 88*(3), 327–342. https://doi.org/10.1007/s10649-014-9541-7

Glade, M. (2014). *Individuelle Prozesse der fortschreitenden Schematisierung.* Wiesbaden: Springer Spektrum.

Glade, M., & Prediger, S. (2017). Students' individual schematization pathways – empirical reconstructions for the case of part-of-part determination for fractions. *Educational Studies in Mathematics, 94*(2), 185–203.

Gould, R. (2004). Variability: One statistician's view. *Statistics Education Research Journal, 3*(2), 7–16.

Gravemeijer, K. (1999). How Emergent Models May Foster the Constitution of Formal Mathematics. *Mathematical Thinking and Learning, 1*(2), 155–177. https://doi.org/10.1207/s15327833mtl0102_4

Gravemeijer, K., & Cobb, P. (2006). Design Research from the Learning Design Perspective. In J. van den Akker, K. Gravemeijer, S. McKenney, & N. M. Nieveen (Eds.), *Educational Design Research: The design, development and evaluation of programs, processes and products* (pp. 45–85). London: Routledge.

Greeno, J. G. (1998). The Situativity of Knowing, Learning, and Research. *American Psychologist, 53*(1), 5–26.

Hardy, G. H. (2005). *A Mathematician's Apology.* Alberta: University of Alberta Mathematical Sciences Society. Retrieved from http://www.math.ualberta.ca/mss/

Harradine, A., & Konold, C. (2006). How Representational Medium Affects the Data Displays Students Make. In A. J. Rossman & B. Chance (Eds.), *Working cooperatively in statistics education: Proceedings of the 7th International Conference on the Teaching of Statistics. [CD-ROM].* Voorburg: International Association for Statistical Education.

Heymann, H. W. (2010). *Why teach mathematics? A focus on general education.* Dordrecht: Kluwer.

Hußmann, S., Thiele, J., Hinz, R., Prediger, S., & Ralle, B. (2013). Gegenstandsorientierte Unterrichtsdesigns entwickeln und erforschen. Fachdidaktische Entwicklungsforschung im Dortmunder Modell. In M. Komorek & S. Prediger (Eds.), *Fachdidaktische Forschungen: Vol. 5. Der lange Weg zum Unterrichtsdesign: Zur Begründung und Umsetzung fachdidaktischer Forschungs- und Entwicklungsprogramme* (pp. 25–42). Münster: Waxmann.

Hußmann, S., & Prediger, S. (2016). Specifying and Structuring Mathematical Topics. *Journal für Mathematik-Didaktik, 37*(S1), 33–67. https://doi.org/10.1007/s13138-016-0102-8

Jablonka, E., & Gellert, U. (2007). Mathematisation - Demathematisation. In U. Gellert & E. Jablonka (Eds.), *Mathematisation and demathematisation. Social, philosophical and educational ramifications* (pp. 1–18). Rotterdam: Sense.

Kahneman, D., & Tversky, A. (1979). Prospect Theory: An Analysis of Decision under Risk. *Econometrica, 47*(2), 263. https://doi.org/10.2307/1914185

Kattmann, U., & Gropengießer, H. (1996). Modelliierung der didaktischen Rekonstruktion. In R. Duit & C. v. Rhöneck (Eds.), *IPN: Vol. 151. Lernen in den Naturwissenschaften. Beiträge zu einem Workshop an der Pädagogischen Hochschule Ludwigsburg* (pp. 180–204). Kiel: Institut für die Pädagogik der Naturwissenschaften an der Universität Kiel.

Kattmann, U., Duit, R., Gropengießer, H., & Komorek, M. (1997). Das Modell der Didaktischen Rekonstruktion - ein Rahmen für naturwissenschaftsdidaktische Forschung und Entwicklung. *Zeitschrift für Didaktik der Naturwissenschaften, 3*(3), 3–18.

Klieme, E., Avenarius, H., Blum, W., Döbrich, P., Gruber, H., Prenzel, M.,. . . Vollmer, H. J. (2007). *Zur Entwicklung nationaler Bildungsstandards*. Bonn, Berlin: Bundesministerium für Bildung und Forschung (BMBF).

Konferenz der Kultusminister der Länder in der Bundesrepublik Deutschland (KMK). (2004). *Bildungsstandards im Fach Mathematik für den Mittleren Schulabschluss*. München: Kluwer.

Konold, C., & Pollatsek, A. (2002). Data Analysis as the Search for Signals in Noisy Processes. *Journal for Research in Mathematics Education, 33(4)*, 259. https://doi.org/10.2307/749741

Konold, C., Robinson, A., Khalil, K., Pollatsek, A., Well, A., Wing, R., & Mayr, S. (2002). Students' Use of Modal Clumps to Summarize Data. In B. Phillips (Ed.), *Proceedings of the Sixth International Conference on Teaching Statistics: Developing a statistically literate society. [CD-ROM]*. Voorburg, The Netherlands: International Statistical Institute.

Konold, C., & Miller, C. D. (2011). *Tinkerplots: Dynamic Data exploration*. Emeryville, CA: Key Curriculum Press.

Konold, C., Higgins, T., Russell, S. J., & Khalil, K. (2015). Data seen through Different Lenses. *Educational Studies in Mathematics, 88(3)*, 305–325. https://doi.org/10.1007/s10649-013-9529-8

Kröpfl, B. (2007). *Höhere mathematische Allgemeinbildung am Beispiel von Funktionen*. München, Wien: Profil.

Kröpfl, B., Peschek, W., & Schneider, E. (2000). Stochastik in der Schule: Globale Ideen, lokale Bedeutungen, zentrale Tätigkeiten. *mathematica didactica, 23*(2), 25–27.

Jungwirth, H., & Krummheuer, G. (Eds.). (2006). *Der Blick nach innen: Aspekte der alltäglichen Lebenswelt Mathematikunterricht*. Münster: Waxmann.

Lakatos, I. (1976). *Proofs and refutations: The logic of mathematical discovery*. Cambridge: Cambridge University Press.

Leavy, A. M., & Middleton, J. A. (2011). Elementary and middle grade students' constructions of typicality. *The Journal of Mathematical Behavior, 30*(3), 235–254.

Lehrer, R., & English, L. D. (2018). Introducing Children to Modeling Variability. In D. Ben-Zvi, K. Makar, & J. Garfield (Eds.), *International Handbook of Research in Statistics Education* (pp. 229–260). Cham: Springer.

Lengnink, K. (2010). Vorstellungen bilden: Zwischen Lebenswelt und Mathematik. In T. Leuders, L. Hefendehl-Hebeker, & H.-G. Weigand (Eds.), *Mathemagische Momente* (pp. 120–129). Berlin: Cornelsen.

Lengnink, K. (2013). Prozesse beim Mathematiklernen initiieren und begleiten - vom Wert des Intersubjektiven. In M. Rathgeb, M. Helmerich, R. Krömer, K. Lengnink, & G. Nickel (Eds.), *Mathematik im Prozess: Philosophische, Historische und Didaktische Perspektiven* (pp. 211–223). Dordrecht: Springer.

Lengnink, K., & Peschek, W. (2001). Das Verhältnis von Alltagsdenken und mathematischem Denken als Inhalt mathematischer Bildung. In K. Lengnink, S. Prediger, & F. Siebel (Eds.), *Darmstädter Schriften zur allgemeinen Wissenschaft: Vol. 2. Mathematik und Mensch. Sichtweisen der allgemeinen Mathematik* (pp. 65–82). Mühltal: Verlag Allgemeine Wissenschaft.

Mac Lane, S. (1986). *Mathematics Form and Function.* New York, NY: Springer

Makar, K. (2014). Young children's explorations of average through informal inferential reasoning. *Educational Studies in Mathematics, 86*(1), 61–78. New York, NY: Springer.

Makar, K. (2016). Developing Young Children's Emergent Inferential Practices in Statistics. *Mathematical Thinking and Learning, 18*(1), 1–24.

Makar, K., & Confrey, J. (2003). Clumps, chunks, and spread out: Secondary preservice teachers' reasoning about variation. In C. Lee (Ed.), *Proceedings of the Third International Research Forum on Statistical Reasoning, Thinking, and Literacy (SRTL-3).*

Makar, K., & Confrey, J. (2005). Variation talk: Articulating meaning in statistics. *Statistics Education Research Journal, 4*(1), 27–54.

Makar, K., & Rubin, A. (2009). A Framework for Thinking about Informal Statistical Inference. *Statistics Education Research Journal, 8*(1), 82–105.

Makar, K., Bakker, A., & Ben-Zvi, D. (2011). The Reasoning Behind Informal Statistical Inference. *Mathematical Thinking and Learning, 13*(1-2), 152–173.

Makar, K., & Rubin, A. (2018). Learning About Statistical Inference. In D. Ben-Zvi, K. Makar, & J. Garfield (Eds.), *International Handbook of Research in Statistics Education* (pp. 261–294). Cham: Springer.

Mayring, P. (2000). Qualitative content analysis. *Forum Qualitative Social Sciences, 1*(2). Retrieved from http://www.qualitative-research.net/index.php/fqs/issue/view/28

Mokros, J., & Russell, S. J. (1995). Children's Concepts of Average and Representativeness. *Journal for Research in Mathematics Education, 26*(1), 20. https://doi.org/10.2307/749226

Moore, D. (1990). Uncertainty. In L. A. Steen (Ed.), *On the Shoulders of Giants: New Approaches to Numeracy* (pp. 95–137).

Niss, M., & Jablonka, E. (2014). Mathematical Literacy. In S. Lerman (Ed.), *Encyclopedia of Mathematics Education* (pp. 391–396). Dordrecht: Springer. https://doi.org/10.1007/978-94-007-4978-8_100

Noss, R., & Hoyles, C. (1992). Looking back and looking forward. In C. Hoyles & R. Noss (Eds.), *Learning mathematics and Logo* (pp. 431–468). Cambridge, Mass.: The MIT Press.

Noss, R., Healy, L., & Hoyles, C. (1996). The Construction of Mathematical Meanings: Connecting the Visual with the Symbolic. *Educational Studies in Mathematics, 33*(2), 203–233. https://doi.org/10.1023/A:1002943821419

Noss, R., Hoyles, C., & Pozzi, S. (2002). Abstraction in Expertise: A Study of Nurses' Conceptions of Concentration. *Journal for Research in Mathematics Education, 33*(3), 204. https://doi.org/10.2307/749725

Organisation for Economic Cooperation and Development (OECD) (2017). PISA 2015 Mathematics Framework. In *PISA 2015 Assessment and Analytical Framework: Science, Reading, Mathematic and Financial Literacy*. Paris: OECD.

Paparistodemou, E., & Meletiou-Mavrotheris, M. (2008). Developing young students' informal inference skills in data analysis. *Statistics Education Research Journal, 7*(2), 83–106.

Peschek, W., Prediger, S., & Schneider, E. (2008). Reflektieren und Reflexionswissen im Mathematikunterricht. *PM - Praxis der Mathematik in der Schule, 50*(20), 1–6.

Petocz, P., Reid, A., & Gal, I. (2018). Statistics Education Research. In D. Ben-Zvi, K. Makar, & J. Garfield (Eds.), *International Handbook of Research in Statistics Education* (pp. 71–99). Cham: Springer.

Pfannkuch, M. (2011). The Role of Context in Developing Informal Statistical Inferential Reasoning: A Classroom Study. *Mathematical Thinking and Learning, 13*(1-2), 27–46.

Pfannkuch, M., Arnold, P., & Wild, C. J. (2015). What I See is not Quite the Way it Really is: Students' Emergent Reasoning about Sampling Variability. *Educ Stud Math, 88*, 343–360.

Pfannkuch, M., Budgett, S., & Arnold, P. (2015). Experiment-to-causation inference: Understanding causality in a probabilistic setting. In A. Zieffler & E. Fry (Eds.), *Reasoning about uncertainty: Learning and teaching informal inferential reasoning* (pp. 95–127). Minneapolis, Minnesota: Catalyst.

Plomp, T., & Nieveen, N. (Eds.). (2013). *Educational Design Research - Part A: An Introduction*. Enschede: SLO.

Polya, G. (1945). *How to Solve It: A New Aspect of Mathematical Method. Princeton science library*. Princeton: Princeton University Press.

Porter, T. M. (1995). *Trust in numbers: The pursuit of objectivity in science and public life*. Princeton, N.J.: Princeton University Press.

Pratt, D., & Noss, R. (2002). The Microevolution of Mathematical Knowledge: The Case of Randomness. *Journal of the Learning Sciences, 11*(4), 453–488. https://doi.org/10.1207/S15327809JLS1104_2

Pratt, D., & Noss, R. (2010). Designing for Mathematical Abstraction. *International Journal of Computers for Mathematical Learning, 15*(2), 81–97.

Pratt, D., & Kazak, S. (2018). Research on Uncertainty. In D. Ben-Zvi, K. Makar, & J. Garfield (Eds.), *International Handbook of Research in Statistics Education* (pp. 193–227). Cham: Springer.

Prediger, S. (2004). *Mathematiklernen als interkulturelles Lernen. Mathematikphilosophische, deskriptive, und präskriptive Betrachtungen* (Habilitationsschrift). Universität Klagenfurt.

Prediger, S. (2005a). „Auch will ich Lernprozesse beobachten, um besser Mathematik zu verstehen." - Didaktische Rekonstruktion als mathematikdidaktischer Forschungsansatz zur Restrukturierung von Mathematik. *mathematica didactica, 28*(2), 23–47.

Prediger, S. (2005b). Developing reflectiveness in mathematics classrooms— An aim to be reached in several ways. *ZDM, 37*(3), 250–257.

Prediger, S. (2008). Do You Want Me to Do It with Probability or with My Normal Thinking? Horizontal and Vertical Views on the Formation of Stochastic Conceptions. *IEJME-Mathematics Education, 3*(3), 126–154.

Prediger, S. (2015). Theorien und Theoriebildung in didiaktischer Forschung und Entwicklung. In R. Bruder, L. Hefendehl-Hebeker, B. Schmidt-Thieme, & H.-G. Weigand (Eds.), *Handbuch der Mathematikdidaktik*. Berlin: Springer.

Prediger, S., Link, M., Hinz, R., Hußmann, S., Thiele, J., & Ralle, B. (2012). Lehr-Lernprozesse initiieren und erforschen–fachdidaktische Entwicklungsforschung im Dortmunder Modell. *Mathematischer und Naturwissenschaftlicher Unterricht, 65*(8), 452–457.

Prediger, S., & Zwetzschler, L. (2013). Topic-specific Design Research with a Focus on Learning Processes: The Case of Understanding Algebraic Equivalence in Grade 8. In T. Plomp & N. Nieveen (Eds.), *Educational Design Research – Part A: An Introduction* (pp. 409–423). Enschede: SLO.

Prediger, S., & Schnell, S. (2014). Investigating the Dynamics of Stochastic Learning Processes: A Didactical Research Perspective, Its Methodological and Theoretical Framework, Illustrated for the Case of the Short Term–Long Term Distinction. In E. J. Chernoff & B. Sriraman (Eds.), *Advances in mathematics education. Probabilistic thinking: Presenting plural perspectives* (pp. 533–558). Dordrecht: Springer.

Prediger, S., Gravemeijer, K., & Confrey, J. (2015). Design Research with a Focus on Learning Processes: An Overview on Achievements and Challenges. *ZDM, 47*(6), 877–891. https://doi.org/10.1007/s11858-015-0722-3

Prediger, S., Schnell, S., & Rösike, K.-A. (2016). Design Research with a focus on content-specific professionalization processes: The case of noticing students' potentials. In S. Zehetmeier, B. Rösken-Winter, D. Potari, & M. Ribeiro (Eds.), *Proceedings of the Third ERME Topic Conference on Mathematics Teaching, Resources and Teacher Professional Development* (pp. 96–105). Berlin: Humboldt-Universität zu Berlin / HAL.

Prediger, S., Leuders, T., & Rösken-Winter, B. (2017). Drei-Tetraeder-Modell der gegenstandsbezogenen Professionalisierungsforschung: Fachspezifische Verknüpfung von Design und Forschung. *Jahrbuch der allgemeinen Didaktik, 2017*, 159–177.

Prediger, S., & Zindel, C. (2017). School Academic Language Demands for Understanding Functional Relationships: A Design Research Project on the Role of Language in Reading and Learning. *EURASIA Journal of Mathematics, Science and Technology Education, 13*(7b), 4157–4188.

Quetelet, L.-A.-J. (1994). A Treatise on Man and the Development of His Faculties. *Obesity Research, 2*(1), 72–85. https://doi.org/10.1002/j.1550-8528.1994.tb00047.x

Reading, C., & Shaughnessy, J. M. (2004). Reasoning About Variation. In D. Ben-Zvi & J. Garfield (Eds.), *The Challenge of Developing Statistical Literacy, Reasoning and Thinking* (pp. 201–226). Dordrecht: Springer. https://doi.org/10.1007/1-4020-2278-6_9

Riemeier, T., & Gropengießer, H. (2008). On the Roots of Difficulties in Learning about Cell Division: Process-based analysis of students' conceptual development in teaching experiments. *International Journal of Science Education, 30*(7), 923–939.

Schnell, S. (2013). *Muster und Variabilität erkunden: Konstruktionsprozesse kontextspezifischer Vorstellungen zum Phänomen Zufall*. Zugl.: Dortmund, Technisch Univ., Dissertation, 2013. *Dortmunder Beiträge zur Entwicklung und Erforschung des Mathematikunterrichts: Vol. 14*. Wiesbaden: Springer Spektrum.

Schnell, S., & Büscher, C. (2015). Individual Concepts of Students Comparing Distribution. In K. Krainer & N. Vondrová (Eds.), *Proceedings of the Ninth Congress of the European Society for Research in Mathematics Education* (pp. 754–760). Prague, Czech Republic: Charles University in Prague, Faculty of Education and ERME.

Schumacher, S. (2017). *Lehrerprofessionswissen im Kontext beschreibender Statistik*. Wiesbaden: Springer Spektrum.

Sfard, A. (2008). *Thinking as Communicating*. Cambridge: Cambridge University Press.

Shaugnessy, M. J. (2007). Research on Statistics Learning and Reasoning. In F. K. Lester (Ed.), *Second handbook of research on mathematics teaching and learning* (pp. 957–1009). Charlotte, NC: Information Age.

Skovsmose, O. (1994). *Towards a Philosophy of Critical Mathematics Education. Mathematics education library: Vol. 15*. Dordrecht: Kluwer.

Skovsmose, O. (1998). Linking Mathematics Education and Democracy: Citizenship, Mathematical Archaeology, Mathemacy and Deliberative Interaction. *Zentralblatt für Didaktik der Mathematik (ZDM)*, *30*(6), 195–203.

Skovsmose, O. (2005). Travelling through education: Uncertainty, mathematics, responsibility. Rotterdam: Sense.

Skovsmose, O. (2012). Symbolic Power, Robotting, and Surveilling. *Educational Studies in Mathematics*, *80*(1-2), 119–132.

Stroeve, J., & Shuman, C. (2004). *Historical Arctic and Antarctic Surface Observational Data*, Version 1. NSIDC: National Snow and Ice Data Center. https://doi.org/10.5067/4din375awfio

Tukey, J. W. (1977). *Exploratory data analysis. Addison-Wesley series in behavioral science*. Reading, Mass.: Addison-Wesley.

van den Akker, J. (1999). Principles and Methods of Development Research. In J. van den Akker, R. M. Branch, K. Gustafson, N. Nieveen, & T. Plomp (Eds.), *Design Approaches and Tools in Education and Training* (pp. 1–14). Dordrecht: Springer.

van den Akker, J. J. H., Gravemeijer, K., McKenny, S., & Nieven, N. (Eds.). (2011). *Educational design research*. London: Routledge.

van den Heuvel-Panhuizen, M. (2003). The Didactical Use of Models in Realistic Mathematics Education: An Example from a Longitudinal Trajectory on Percentage. *Educ Stud Math*, *54*, 9–35.

van den Heuvel-Panhuizen, M. (2001). Realistic mathematics education in the Netherlands. In J. Anghileri (Ed.), *Principles and practices in arithmetic teaching. Innovative approaches for the primary classroom* (pp. 49–63). Buckingham: Open University Press.

Vergnaud, G. (1996). The Theory of Conceptual Fields. In L. P. Steffe (Ed.), *Theories of mathematical learning* (pp. 219–239). Mahwah, N.J.: Erlbaum.

Vergnaud, G. (1998). A Comprehensive Theory of Representation for Mathematics Education. *Journal of Mathematical Behavior*, *17*(2), 167–181.

Vergnaud, G. (2009). The Theory of Conceptual Fields. *Human Development*, *52*, 83–94.

Wagenschein, M. (2010). *Verstehen lehren: Genetisch, sokratisch, exemplarisch* (5. Aufl.). *Beltz-Taschenbuch: Vol. 22*. Weinheim: Beltz.

Watson, J. M. (2007). The Role of Cognitive Conflict in Developing Students' Understanding of Average. *Educational Studies in Mathematics*, *65*(1), 21–47.

Watson, J., Fitzallen, N., Fielding-Wells, J., & Madden, S. (2018). The Practice of Statistics. In D. Ben-Zvi, K. Makar, & J. Garfield (Eds.), *International Handbook of Research in Statistics Education* (pp. 105–137). Cham: Springer.

White, P., & Gorard, S. (2017). Against inferential statistics: How and why current statistics teaching gets it wrong. *Statistics Education Research Journal*, *16*(1), 55–65.

220

References

Wild, C. J. (2006). The concept of distribution. *Statistics Education Research Journal*, 5(2), 10–26.

Wild, C. J., & Pfannkuch, M. (1999). Statistical Thinking in Empirical Enquiry. *International Statistical Review*, 67(3), 223–248.

Wild, C. J., Utts, J. M., & Horton, N. J. (2018). What is Statistics? In D. Ben-Zvi, K. Makar, & J. Garfield (Eds.), *International Handbook of Research in Statistics Education* (pp. 5–36). Cham: Springer.

Wille, R. (1981). Versuche der Restrukturierung von Mathematik am Beispiel der Grundvorlesung „Lineare Algebra". In B. Artmann (Ed.), *Beiträge zum Mathematikunterricht* (pp. 102-112). Hannover: Schroedel.

Wille, R. (1988). Allgemeine Wissenschaft als Wissenschaft für die Allgemeinheit. In H. Böhme & H.-J. Gamm (Eds.), *Verantwortung in der Wissenschaft* (pp. 159-176). Darmstadt: Technische Hochschule. Reprint in Conceptus – Zeitschrift für Philosophie, 60 (1989), 117-128.

Wille, R. (1995). „Allgemeine Mathematik" als Bildungskonzept für die Schule. In R. Biehler (Ed.), *Mathematik allgemeinbildend unterrichten. Impulse für Lehrerbildung und Schule* (2nd ed., Vol. 21, pp. 41 55). Köln: Aulis.

Wille, R. (2000). Bildung und Mathematik. *Mathematische Semesterberichte*, 47, 11–25.

Wille, R. (2008). Generalistic mathematics as mathematics for the general public. In G. Dorfer (Ed.), *Contributions to general algebra: Vol. 18. Proceedings of the 73th Workshop on General Algebra. "73. Arbeitstagung Allgemeine Algebra" ; 22nd Conference of Young Algebraists, Alps-Adriatic-University of Klagenfurt, February 1 - 4, 2007* (pp. 211–225). Klagenfurt: Heyn.

Winter, H. (1981). Der didaktische Stellenwert des Sachrechnens im Mathematikunterricht der Grund- und Hauptschule. *Paed. Welt*, 35(11), 666–674.

Winter, H. (1996). Mathematikunterricht und Allgemeinbildung. *Mitteilungen der Deutschen Mathematiker-Vereinigung*, 4(2), 35-41. https://doi.org/10.1515/dmvm-1996-0214

Wittgenstein, L. (1953/2008). *Philosophische Untersuchungen*. Frankfurt am Main: Suhrkamp.

Wood, D., Bruner, J. S., & Ross, G. (1976). The role of tutoring in problem solving. *Journal of Child Psychology and Psychiatry*, 17(2), 89–100. https://doi.org/10.1111/j.1469-7610.1976.tb00381.x

Zieffler, A., Garfield, J., Delmas, R., & Reading, C. (2008). A framework to support research on informal inferential reasoning. *Statistics Education Research Journal*, 7(2), 40–58.

Zieffler, A., Garfield, J., & Fry, E. (2018). What is Statistics Education? In D. Ben-Zvi, K. Makar, & J. Garfield (Eds.), *International Handbook of Research in Statistics Education* (pp. 37–70). Cham: Springer.

Ziliak, S. T., & McCloskey, D. N. (2009). *The cult of statistical significance: How the standard error costs us jobs, justice, and lives*. Ann Arbor, Mich.: Univ. of Michigan Press.

Zwetzschler, L. (2015). *Gleichwertigkeit von Termen*. Wiesbaden: Springer Spektrum.

MIX
Papier aus verantwortungsvollen Quellen
Paper from responsible sources
FSC® C105338

FSC
www.fsc.org

If you have any concerns about our products,
you can contact us on
ProductSafety@springernature.com

In case Publisher is established outside the EU,
the EU authorized representative is:
Springer Nature Customer Service Center GmbH
Europaplatz 3, 69115 Heidelberg, Germany

Printed by Libri Plureos GmbH
in Hamburg, Germany